SpringerBriefs in Geography

SpringerBriefs in Geography presents concise summaries of cutting-edge research and practical applications across the fields of physical, environmental and human geography. It publishes compact refereed monographs under the editorial supervision of an international advisory board with the aim to publish 8 to 12 weeks after acceptance. Volumes are compact, 50 to 125 pages, with a clear focus. The series covers a range of content from professional to academic such as: timely reports of state-of-the art analytical techniques, bridges between new research results, snapshots of hot and/or emerging topics, elaborated thesis, literature reviews, and in-depth case studies.

The scope of the series spans the entire field of geography, with a view to significantly advance research. The character of the series is international and multidisciplinary and will include research areas such as: GIS/cartography, remote sensing, geographical education, geospatial analysis, techniques and modeling, landscape/regional and urban planning, economic geography, housing and the built environment, and quantitative geography. Volumes in this series may analyze past, present and/or future trends, as well as their determinants and consequences. Both solicited and unsolicited manuscripts are considered for publication in this series.

SpringerBriefs in Geography will be of interest to a wide range of individuals with interests in physical, environmental and human geography as well as for researchers from allied disciplines.

More information about this series at http://www.springer.com/series/10050

Zaheer Allam

Theology and Urban Sustainability

Springer

Zaheer Allam
The Port Louis Development Initiative
Port Louis, Mauritius

ISSN 2211-4165 ISSN 2211-4173 (electronic)
SpringerBriefs in Geography
ISBN 978-3-030-29672-8 ISBN 978-3-030-29673-5 (eBook)
https://doi.org/10.1007/978-3-030-29673-5

This Springer imprint is published by the registered company Springer Nature Switzerland AG
The registered company address is: Gewerbestrasse 11, 6330 Cham, Switzerland

Foreword

When Yuval Harari spoke about religion, people were surprised when he took a bold stance and equated it to fake news. However, when we hear the term today, we immediately think of social media, but we are wrong. They are much older than Facebook. Today, more than ever, we are aware that there is zero scientific evidence that Eve was tempted by the snake, that the creator is unhappy when a Brahmin marries an Untouchable, that all infidels will burn in hell. It is understandable that one can easily be upset that religion is being associated with such a trivial matter like fake news. Indeed, we cannot ignore that billions of people have believed in religious tenets for centuries and that religion has inspired people to create wonderful monuments dedicated to their creator(s), to build hospitals, schools, and universities.

Whole cities have been planned and organized through the association of two powers: religion and monarchy. This duality of power has shaped cities from the days of ancient Egypt, Persia, and Mesopotamia and led the Crusaders, including the Templars, into Holy War, creating grand fortified castles in cities as distant as Aleppo in Syria. Moorish Spain has seen magnificent churches converted into mosques and back, just like Byzance turned into Constantinople and then into Istanbul. But it has also inspired men to create armies, prisons, and to burn down cities to ashes. In the name of religion, pre-Colombian civilizations have been wiped. The Dutch Reformed Church (NGK) of South Africa found justification channeling a select group of "chosen people" to segregated zoning systems and redefining residential areas. Today this is unjust, unfair, and illogical. Yesterday it was another story.

But religion has not been the monopoly of gods. Modern times led to the invention of new forms of religion. Communism has created Kolkhozes and Gulags, fascism spread its own architecture as far as Eritrea. And today, in our digital age, ICT is trying to protect safe cities against invisible Holy Wars. The dollar note says "In God we trust"—which one? we are unsure. Coca Cola has managed to associate itself with youth, health, and sports. And above all, there is one new religion that has surpassed all other beliefs: the worship paid to the CAR, a dying industry's reigning majesty. The CAR which promised billions of people the joys of life *à la campagne* while working and enjoying all the benefits of the city. Much of today's urban sprawl is attributable to its invention. The CAR, creating suburbs and huge shopping malls in the middle of nowhere, the latter at the center of an intoxicating ballet choreographed for thousands of cars. Just like shopping malls, cars today surround places of worship, as people flock to consume some sort of abstract form of guidance.

Zaheer's book goes deep into these notions. It highlights the fact that religion makes it possible to organize people effectively and to structure the way they live through consensual social agreements. We now long to see a sequel to this book, one that will go deep inside history to better project us forward into the future. A future that will explore other spheres, that will bind people together as successfully as any god or holy book.

Port Louis, Mauritius

Gaetan Siew
Governing Council, UN Habitat
Past President, International
Union of Architects

Preface

While my interest ranges primarily in the study, Smart Cities' Sustainability and Resilience, I often found myself pondering on the origins of many of our practices, those we take for granted and do not dispute. Often, they are linked to and shaped by politics and societal shifts. However, going deeper we notice that many of our doings (or wrongdoings) can be associated with some notion of religion.

We cannot deny the powerful influence that religion has had on our societies. We all marvel at the great architectural beauty of places of worships, we are wonderstruck by the way old cities were planned and designed in the minutest of detail, all aimed at providing a recollected setting for people to meditate in peace. This profound manifestation of faith and the efforts entailed to provide platforms for people to interact with higher power(s) can lead us to think in retrospect. Taking stock of all the energy invested and often the linear thinking (as religion can indeed blind people), we can wonder whether all this is actually sustainable. Are we acting blindly on the basis of pledges meant to be redeemed in a vague future or is there some abstract rationality that can be retrieved?

The nature of my work as an Urban Strategist afforded me to travel quite often with Gaëtan Siew, Past President of the International Union of Architects, for who I work for, and my mentor. During these trips to distant places, we got to see first-hand how expensive places of worship can be erected right in the middle of places of misery. We have also observed how irrational people can act when it comes to defending religious beliefs, how religion can be used as an excuse to support segregation and cause distinct cities for the rich and poor. We then ask: if in the name of religion, people can act with so much discord, how come is it also synonymous to peace for many? I am thus happy that Gaëtan gracefully accepted to write the foreword of this book, as it is fitting that this discussion starts and closes with him.

Interestingly, while pursuing this discussion, we noted that there are overarching principles that can be seen emerging in comparative religious studies. Maybe this is why the quest for enlightenment guides people to interact with interfaith groups. This quest can then be found to support a definition of spirituality, attained by the understanding that there are commonalities between various religious beliefs. This encourages acceptability of differences.

So, how do we approach this contradictory nature of religion, the duality of darkness and light? And how does this relate to the urban form?

This book explores this conundrum and advances some theories unveiling the damages that religion may have caused—but which are overlooked, and also explores how some religious principles, interpreted by others, like Christopher Alexander, Nikos Salingaros, and Michael Mehaffy can be helpful in guiding us towards how to build a truly cohesive world.

Looking forward, this book also dwells into the thematic of the exponential rise of data, how it can be a catalyst for the decline of religion and how one may wonder if data in itself can be seen as a new religion; if not a new God... This will also mean that we are perceiving a shift in the very nature of religion and of our relation thereto. In this case, data is seen to work for us, instead of us working for a higher power. So where does this place us in the abstract hierarchy of power? This raises philosophical questions, many of which I am not equipped to answer. We must nevertheless be prepared to ask them, as we live in a time of unprecedented challenges, the more so as the survival of humankind is at stake. I see the irony here, as most religious texts predicted some kind of apocalypse looming around the corner.

However, we need more than prayers, we need action and, most importantly, remediation.

Port Louis, Mauritius Zaheer Allam

Acknowledgements This work was borne from conversations with many people and I am indebted to Gaetan Siew, Gilbert Ahnee, Sameerah Aumjaud, Nawez Atawoo, Khatleen Minerve, Houriiyah Tegally, Abdur Rahmaan Denmamode, Peter Newman, Nikos Salingaros, Michael Mehaffy and Dean Saitta for never-ending conversations that continue to shape my understanding on the way our society structures our world across time.

Contents

About the Author

Zaheer Allam is a holder of a Ph.D. from Curtin University (Australia), an MBA from Anglia Ruskin University (UK) and a Bachelor of Applied Science in Architectural Science from Curtin University (Australia). Based in Mauritius, he works as an Urban Strategist for The Port Louis Development Initiative (PLDI), the Global Creative Leadership Initiative and consults on a number of projects on the thematic of Smart Cities across the African Continent and on strategies dwelling in the increasing role of technology in Culture and the Society. Zaheer is also the African Representative of the International Society of Biourbanism (ISB), member of the Advisory Circle of the International Federation of Landscape Architects (IFLA), and a member of a number of other international bodies. For his contributions to society, he was elevated, by the President of Mauritius, to the rank of Officer of the Order of the Star and Key of the Indian Ocean (OSK), the highest distinct order of Merit in Mauritius. He is also a recipient of a number of other awards and writes extensively on the thematic of sustainability and resilience in cities.

Photo credit to Yatish Ramdharrysing.

Chapter 1
Religion and Urban Planning

Abstract Religious influences on urbanism are not only manifested through how places of worship have shaped cities across time but also encompass sociological, cultural, and political dimensions. This is witnessed through the grand monuments that stood the test of time, or through how the relationship of religion and politics has had a long story. Even though modern society is more educated, the role of religion is still having an impact on contemporary urban form and on society at large. This chapter explores how this phenomenon occurs in the discourse of increasing urbanization coupled with the impacts of climate change. In view of the challenges faced by cities, this chapter further ponders if religious dogmas are still applicable and current, and thus serving society just as well as religious bodies.

Keywords Religion · Cities · Politics · Society · Psychology · Faith · Urban planning · Urban morphology

1.1 Introduction

Since time immemorial, religion has played significant roles in the shaping of societies; both in terms of beliefs and behavior (Powell and Clarke 2012). One characteristic that stands out is the infusion of beliefs and values that one ought to ascribe to (Kirchmaier et al. 2018). In some quarters, religion is viewed as a belief system that originated from the cooperation and commitment of individuals or groups of people, and this in pursuit of some supernatural entities and forces (Mercier et al. 2018). Others perceive it as a by-product of cognitive functions and activities that are non-religious by nature but have the power to impact on people in their search for emotional and psychological wellbeing (Boyer 2003).

One area where religion greatly impacted society is the emphasis on segregation through the bias of what is commonly known as castes. Initially, castes had a significance on how different religious activities, such worshipping, were conducted and it was noticed that those in higher castes had higher degree of responsibilities than their religious counterparts following the same set of beliefs (Mosse 2018; Sankaran et al. 2017). This religious segmentation led to different classes of people commanding varying responsibilities, and respect, depending on the role they played in the

Z. Allam, *Theology and Urban Sustainability*, SpringerBriefs in Geography,
https://doi.org/10.1007/978-3-030-29673-5_1

group (Hage and Posner 2015). As religion evolved from the simple groupings of people with shared belief systems to accommodate an increasing demographic, their influence on society equally increased. For instance, Nath (2015) posits that religious groups have evolved from the worshipping of deities to engaging in different kind of activities including the support of the education system, contributing in health-care, caring for the need of the poor and the homeless, and even to the extent of participating in matters of national interests. These played significant roles leading to increased investments in basic and primary sectors resulting in how areas are view by the public; resulting to a sense of responsibility from policymakers in regard to behaving ethically (Dallahite et al. 2018). However, noting the power and respect religious leaders yield, there have been notable cases of corruption and overuse of power and influence to increase political control (Shadabi 2013), as it has been seen in many cases, especially in relationship dynamics between religion and governance (Aldashev and Platteau 2014).

There are numerous ways how those belief systems impacted cities (Clarke 2011; Kauffman 2010). One of which is the emphasis on families, and the need for decent housing (Nassar and Elsayed 2017; Dean and Pryce 2017). However, this has seen considerable change over the years as in earlier days, it is noted that most societies were made up of households that comprised of large number of peoples, as in line with most religious teachings. With the evolving of religious beliefs, their view on matrimonial and family institution changed, leading to the development of nuclear family units; which can be argued as a proponent of urban sprawl. The concept of maintaining small family units is a positive influence since it is easier to provide basic urban needs like shelter, food, education, healthcare and clothing among many others. But on the other hand, with the increasing challenges like scarcity of resources and environmental degradation, the small family unit concept is deemed by some quarters as unsustainability. This is further explored in Chap. 2.

In urban centers, the footprints of religion can be seen in almost every ancient city and central places that have existed (Aldashev and Platteau 2014). Evidences of those can be seen around the world and are manifested in the scale and considerable amounts of resources invested in erecting grand monuments such as the great pyramids of Giza (Fig. 1.1), the magnificent temples in Old Bagan (Fig. 1.2), the Chidambaram Thillai Nataraja temple (Fig. 1.3), and Cathedral of Saint Mary of the See (Fig. 1.4), among others.

Fernández-Götz (2018) affirms this proposition and expounds that religion and political aspects has had significant influences on cities and their fabrics, especially noting that they two functions command grandeur and power, and a majority of ancient rulers had their roots either on religion or political backgrounds. Even in modern days, these two functions have a great influence on cities, and there is tangible evidence that most cities are developed at the backdrop of religious and political dimensions. Day (2017) adds that most cities are designed to align with religious principles as a backdrop and this is made to impact on followers of such faiths, such that they can frequently visit them for different religious practices. Goh and van der Veer (2016) explain that this trend is fuelled by the desire to fullfill the perceived psychological need to be closer to God; hence, even the infrastructure of the city

Fig. 1.1 Pyramids of Giza. Photo by Pradeep Gopal

Fig. 1.2 Old Bagan, Myanmar. Photo by Sven Scheuermeier

Fig. 1.3 Chidambaram Thillai Nataraja temple, Chidambaram, India. Photo by Naveen Raj Dhanapal

Fig. 1.4 Cathedral of Saint Mary of the See, Seville, Spain. Photo by Henrique Ferreira

Fig. 1.5 The Vatican City. Photo by Christopher Czermak

is made to symbolically and materially mirror the beliefs of the religious group in power. Prominent examples include the Vatican City (Fig. 1.5), which is the center for the Catholic religion and Mecca (Fig. 1.6), from the Kingdom of Saudi Arabia, is marked as the origin place of the Islamic Faith.

In these cities, almost every urban fabric is fashioned and designed to align with the religious dictates. Zhang and Wei (2017) further share that the city of Lhasa, and societies, in Tibet, are also heavily influenced by religion, and those are now heavily marketed as religious tourism (Kang 2009). The link between religion and economy is furthermore apparent and further discussed in (McCleary and Barro 2006; Tomalin 2018).

Despite the heavy investments and influence religion has had on the society and on the physical places like cities and urban areas, a majority of the modern population is observed to leave behind religious beliefs in favor of spirituality. Overstreet (2010) explains that this happens when individuals recognize and believe in God and in its different forms, but do not base their beliefs in God to any organized religion. This is now commonly defined as being "spiritual, but not religious" (Carey 2018). These trends have been observed to have significant implications on religion and Rinallo and Oliver (2019) confirm that the impacts of spiritualization are already being felt in how people are reacting to modern trends like technology, secularization, consumerism, relationships, leadership and governance among other issues. The authors argue that spiritualization is challenging established religious practices and recognizing contemporary issues that do not necessarily augur well with religion.

Fig. 1.6 Mecca, Medina, Saudi Arabia. Photo by Adli Wahid

1.2 The Place of Worship and Cities

The role of religion as a tool for shaping cities is well documented and apparent through the grandeur and omnipresence of places of worship in city places, and more so in cities known for their role in politics and economy at a larger local and regional scale. Interestingly, from a historical perspective, it is known that the first cities developed around places of worship and sacred grounds. Mazumdar and Mazumdar (2013) explain that, unlike in the modern days, ancient urban planners and developers had to consult with religious leaders, who, in most cases served, de facto, as the political leaders. The genesis of such practices is partly credited to the fact that religion played a key role in the establishment of those cities, and thus an influenced remained on how to plan in accordance to their needs and wishes so as to attract an even larger population; hence, larger economy (Bai and Kung 2015). The presence of places of worship in this arena would not only help to remind people to abide by the established set of belief systems, but also act as a symbol of power. For instance, Mumford (1961) highlights that the cathedral was a focal point of ancient European cities and its influence spanned in its entire cityscape, and this is still seen today. For example, the skyline of the old city of Rome, established since 753 BCE (Mark 2009) can still be seen as being dominated by places of worships and symbols associated to them. In a similar fashion, the designs of neighborhoods, streets, roads, green spaces, and recreation sites have been noted as being directly influenced by the religion (Boeck 2014).

On a different aspect, besides the physical presence of religious buildings and through physical forms, including that of ornaments and crafts, religious beliefs also influenced how cities were planned and developed through design philosophies. Lynch (1981) shares an example of ancient Easter cities, where its design was inspired after leaders had consulted the cosmos concerning such things like harmony, preservation of good over evil and even the urban morphology through the specific localization of urban elements and components (Mazumdar and Mazumdar 2013). Other notable examples include the temple city of Madurai in India (Kumar and Aruna 2018), the Sinhalesi Buddhist city in Sri Lanka (Lakpura 2018), and the Feng Shui in China (China Travel 2018). Shokoohy (1983) further notes that places of worship influenced the evolution and living patterns of cities. Caravanserais found in the Silk Roads; from Turkey to China, and also in parts of North Africa (UNESCO 2010), for example, represent this perfectly as those were frequently erected near places of worship to first, facilitate travelers to worship and second, to capitalize on their presence for trade, provide a place for people to interact, eat, and other such things (UNESCO 2010).

While such practice may seem strange today, the role of religion in shaping the cities is still live and prominent; though not manifested in mystic form. For instance, different cities witnessing migration flows, especially in U.S and Europe (Pedersen and Rytter 2017; Cesari 2006; Hirschman 2007), are seeing a clash in culture as varying religious beliefs are still portrayed as how different styles of living, dressing, and eating are seen, as well as the identity and choice of neighborhoods emerge (Hirschman 2007). Interestingly, even response to issues like the environmental stewardship and law vary (Koehrsen 2017; Altmann et al. 2012), which are made to impact greatly on how segregated neighborhoods are designed, and how they interact with other neighborhoods. However, there is an interesting emerging feature: Economics, as places of worship are seen to attract a large crowd, and thus provide real estate commercial viability, spurring positively on property values (Proppe 2015; Lehrer 2004).

The flow of large crowds of people is not only of interest to business-oriented individuals and commercial organizations but also the religious groups dominant in given areas. It has been found that most places of worship are now being considered as commercial entities and religious groups have invested heavily to ensure a sustained, or even an increasing, movement of people; hence, rendering commercial activities related to the place of worship even more rewarding. Rashid (2018) however notes that not everyone that frequents the place of worship is motivated by the actual act of worshipping. This particular notion is what the management of the commercially oriented departments, active in places of worship capitalizes on. Going further, Kislali et al. (2016) posit that places of worship even go to the extent of positioning themselves as tourism sites, and this is now current in the form of religious tourism. This has become such a lucrative undertaking that even religious symbols and signs are being protected under the trademarks laws (Liu 2018). However, the argument of the various religious spokesperson is that this is done to preserve religious cultures from being misused, while on the other side it is noted that some places of worship use specific trademark symbols as a way to market their sites. Building on further, religious groups usually market different products in the form of books, sacred ornaments, entertainment products like songs and movies and religious clothing that are aimed at differentiating them, but most importantly stand as a strong source of revenue (Terzidou et al. 2018). Interestingly, to the search to drive increased profits, it is also common for religious groups to invest in the hotel and hospitality industry, transport sector, media, entertainment industry, education and health sectors (Cranenburgh et al. 2014). This trend of commercialization can be seen to expand from the place of worship and adopted by external (often with no official association) vendors who make use religious symbols to better target people, as shown in Fig. 1.7.

Though commercialization of places of worship and faiths may be innocent to some groups, in its very essence, religion ought to be separated from commerce and other undertakings that may seem to conflict the very purpose of religion in the society. Religion is sought to uphold its original purpose of acting as a bridge between the world of the living and the dead (Koenig 2012; Harding et al. 2005). Many believe that it should preserve its sacred role of defining what is ethical and unethical and serve as a guide to other institutions like the family, the political class among other groups (Arikan and Bloom 2018; DeHanas and Shterin 2018). Through this end, the profit-driven agenda further plunges religion into confusion as in many cases, it has been observed to engage in corruption, abuse of human rights, wars, and human-made famine and calamities. Einstein (2008) argues that all those contribute to the loss of credibility by places of worship and helps in promoting a more secular society; thus, diluting the value of religion and its role in the society.

Fig. 1.7 Use of Virgin Mary to help in the sale of wine in Berlin, Germany. Photo by Alina Sofia

1.3 Religion, Society and Psychology

The way people behave has also been heavily influenced by religion. Through religions, principles like love, justice, industriousness, environmental consciousness, and respects among many others are emphasized. There is evidence that different religious groups support the above ethos, contributing to a more ethically inclined society (Laeheem 2018). For instance, Kirchmaier et al. (2018) found out that among the Dutch population, people who subscribed to religious teachings abhorred unethical behaviors like fiscal evasion, bribery, and corruption, and were more willing to help the society through various means, including volunteering. Buckingham (2012) adds that religion promotes such things like law-abiding behaviors and fosters what she calls "moral self-government". Even in terms of humanitarian works, religion adherents globally have been found to be far more committed than their secular counterparts (Markham 2012; Mylek and Nel 2010).

The practice of delegating responsibilities to people in different spheres is an old practice that has been perceived to have significance in ensuring law and order and ensure that morality is observed. Sullivan (1994) expresses that simple action of raising a sacred item like a religious book, cross or any other religious artifacts has a great impact on the psychology of the person, especially with the promise of negative consequence once the oath is breached. Leigh and Hambler (2014) explain that this happens as most religions provide guidelines, norms, and values on different issues on how people are to behave in different spheres of life, especially to maintain social order, and failure to observe them is perceived as being detrimental to society and invite negative consequences upon oneself. Aroney (2018) further explains that in relation to religion and politics, the religious oath that one takes prior to taking office acts as a guarantee that the general public will be served by the officeholder since the oath is perceived to be witnessed by the divine; an external supreme being with the power to provide judgment in case the oath is violated.

In line to the above, different religion teaches against such practices like those related to nepotism, abuse of office, sabotage, taking unfair advantage on things like price and weights and economic crimes like corruption, among many others. On the place of such vices, religion advocates for meekness, industriousness, honesty and stewardship among other virtues. In the words of Leigh and Hambler (2014), ethical behaviors are triggered in religious believers by the conviction that supernatural beings have the potential to see everything people do and this self-examination helps in causing refrain from any economic or social undertaking that would be detrimental to one's consciousness (Atkinson and Bourrat 2011). This observation prompted Zuhaira and Ye-zhuang (2017) to conclude that cities, where religious practices are cherished, tend to be run more efficiently, and experience increased economic benefits relative to those in a secularized environment or those run by non-religious leaders.

Besides the impacts on behavior, religious "training" is perceived to have psychological benefits on its adherents. Fenelon and Danielsen (2016) opine that individuals who are affiliated to a given religion tend to experience better health and wellbeing than their counterparts who were once affiliated, but with time, denounce their religiosity. He argues that regular attendance to places of worship does not only nourish the spirit but has been reported to positively impact on the health of those who frequent them. In reality, this feeling is evoked through the religious bias of providing safety and social support even after events like divorce, bereavement, illness and unemployed among other such like. Similarly, the promise of a better life once an individual depart to the next life is said to contribute to the perceived wellbeing (Brown and Tierney 2009). Despite this conviction by religious people, the discourse is contested and is often equated to being biased as it lacks scientific evidence to support the claim. For instance, the findings by Fenelon and Danielsen (2016) of perceived wellbeing is only reported in adults. This claim is supported by Kosher and Ben-Arieh (2017) who found out that among children, there was no evidence of perceived wellbeing after attending places of worship.

The influence of religion on the life of an individual has a bearing on how such a person acts when they are mandated to design a city or its fabrics (Allam 2019). In particular, due to fear of going against the religious tenets on different issues appertaining to their mandate; their decision making becomes impaired. When this happens, the resulting consequences are making irrational, or biased, decisions that are highly influenced by the dominant religion in the area. This is true despite the fact that such cities may be cosmopolitans with individuals who subscribes to different religions, and thus, those from minority religions have to contend with city fabrics that are not representative of what they believe. Similarly, the designs being dictated may not be coherent with modern trends like use of ICT and other advanced technologies; hence, cannot compete with its peer or address modern urban challenges like population increase, scarcity of land and urban sprawl among many other issues. A case in point of cities influenced by religion include the cities of Eindhoven and Roermond located in the southern part of Netherlands that Janssen (2016) reports that were highly influenced by the Catholicism in their designing, and the same is visible in every aspect and fabric in them. For instance, there are numerous, towering places of worship in public places and parochial complexes in distinct locations of the city. Those are surrounded by a number of associated activities likes religious schools, commerce, and others and even impact on the demographics in surrounding residential estates in these cities. Another example is the city of Birjand, in Iran, as explained by Fakhimzade et al. (2016), which is highly influenced by the Islamic culture with structures like mosques and markets selling religious artifacts being among the prominent pointers. Others include the Lhasa, a perceived sacred city in Tibet that is majorly influenced by Buddhism in every aspect (Zhang and Wei 2017). The Vatican City in Rome and Mecca city in the Kingdom of Saudi Arabia also adhere to this list. They serve as headquarters of Catholicism and Islam respectively and are also highly influenced by their respective religion such that each design has to be approved by the religious leaders before they are implemented.

In light of the above, city planners and designers are confronted with the challenge of either adhering to the ethics and morality prescribed by the dominant religion in the city or to design a working city, that incorporate modern trends even though they may sometime negate some of the religious principles.

1.4 God, State, and Law

In the review of ancient cities, it that most of political leaders happened to be the de facto leaders of the dominant religion in the state; highlighting the role of religion in state matters has always been present (Spikard 2013; Ahmed 2017). They yielded so much power such that their input in the legislation of new laws were highly regarded and observed. For instance, the Pope leads Christianity globally from the Vatican City and equally serves as its head of state; thus acting as the de facto leader (Reese 1998). Other states like the Kingdom of Saudi Arabia have been led by monarchy who holds both political and religious offices and the legal system of the country is purely based on Islamic Shariah law. In other cases, some states like Malawi (Quashigah 2014), Iran, Israel, and United Arab Emirates have it outright spelled in their constitution that they are religious states with a dominant religion being the one being followed in these respective countries: Islam (Trimble 2019). By so doing, as Quashigah (2014) expressly shares that the population, regardless of their religious affiliation, are forced to adhere to laws and policies that are based on specific dominant religions. In Mauritania, for instance, the law states that one can only become the head of the state if they are from the Islamic faith (Constitution of the Republic Mauritania Islamic 2006). Besides the proclamation of religious states, a majority of countries have the preamble of their constitution acknowledging the supremacy of a supreme being (ProCon 2008; Orgad 2010), despite the fact that there are individuals who are atheists, and others polytheists.

With religion being part and parcel of many states and governing laws, most are in one way or the other inclined toward specific religion (Nieuwenhuis 2012); hence leading to the creation of social disparities and even in fostering emerging conflicts between religion and state. For instance, the practice of marriage which is recognized in law in different countries is highly influenced by religion, even though if the state is the only authorized body to approve those. This is affirmed by Weller (2015) who argues that most countries, though with power to impose legal and social frameworks, encourage religious marriages. For this reason, though secularism is gaining roots in different places, some marriage practices like gay marriages are not easily accepted and are criticized, rejected and denounced in most countries, especially those with religious inclination (Seetharaman 2017).

In the political sphere, there are numerous practices like election, running of governments and distribution of resources among many others have been seen to be influenced by religion (Oboh 2017; Theodorou 2014). For instance, in some countries such as Vatican City, Afghanistan and Pakistan (Aspinall 2018), due to strict religious rules, women still find it hard to take part in voting processes, and in others, there is are observed inequalities and difficulties for them to hold any political position (UN Women 2018). Due to those influences, Ottuh et al. (2014) opine that religious principles and their related ethos can also be made to contribute greatly in advocating for the inequitable distribution of resources and may further enhance poverty.

While those are seen disconnected to the urban realm, there are considerable impacts to the urban form. In respect to family constitution, for example, different religious groups advocate for monogamous family units; promoting the need for individual households which is collectively designed to provide shelter, education, healthcare and decent education among others (Adsera 2006; Coltrane and Collins 2001; Joo 2015; Olsen 2009). However the provision of those stresses the urban fabric and surrounding urban resources and in some instances, they have been found to contribute to issues like land tussles in urban areas, urban sprawl and environmental degradation, and to further impact on issues like healthcare, education, employment, security, inclusivity and equality among others (Nassar and Elsayed 2017; Koehrsen 2017; Pinter et al. 2016).

The involvement of religion in state matters and politics render unproductive arenas and prevent urban areas to perform effectively. This is because there is a long history of perceived corruption and nepotism; favoring family members and cronies of those in power, while the rest of the people are left to scramble for few job opportunities. Quashigah (2014) explains that such practices have the potential to taint the image of the city; thus discouraging meaningful investments that are paramount for the economic growth of a city. Rieger (2013) argue that some religiously motivated practices promote individualism, hence, issues like timely and efficient provision of services, especially to those lower in the economic status ladder are not given much attention. There have been numerous manifestations asking for the separation of religion and power, but also those that continue to reiterate religious associations and coupling them with political agendas. Figure 1.8 depicts this.

On a societal level, some practices are deemed outdated since they do not augur well with the secularizing of the society. It has been observed that a majority of people

Fig. 1.8 Rally at the White House, Washington, United States. Photo by Melany Rochester

are demanding relaxation of rules such as those that define and guide marriages. They advocate for an open society where they are not discriminated on the basis of their sexual orientation or gender (Seetharaman 2017). This is true as there is evidence of spirited campaign on legalization of same-sex marriages in different parts of the globe. On the same line, most societies are pushing for equal opportunities for both men and women, and they argue that women also have the potential to contribute positively in the growth of their cities, country, region and to the wider global community.

1.5 Faith, Spirituality and Urban Morphology

Interestingly, despite the reign of religion for over thousands of years, its influence is decreasing as people call for the contextualization of religious texts and decree that have influenced the states and societies (Pew Research Center 2015). From the peoples' perspective, those principles are no longer relevant to address present time issues, especially noting that numerous changes like the use of technologies, change in the environment, land use, government and leadership styles and the role of education in the society among many other things (Pew Research Centre 2015; Narayanan 2013; Rakodi 2012). One of the areas where religious influence and dictate is seen to be no longer tenable is in the leadership front where, in most countries, it was illegal for people like women, people with disabilities and those from a different religious group to be seen in a position of power and influence (Haynes 2017). As discussed above, following the Islamic influence, women were not permitted to participate in voting in most Islamic states, especially those of Middle East. This is also true in Vatican City, which is under the influence of Catholic denomination. At the present, due to pressures from global community, arguing mostly from secularized perspective, voting rights have been granted in almost all countries except in Vatican City and others like Somali (Llavador 2017). Women and other marginalized groups in society have also been observed to hold political offices, which was not the case in most countries (Kumar 2017).

Religious decrees and text on the constitution of family, marriage and child-bearing have faced considerable pressure as they are no longer seen to be able, by a larger percentage, to withstand the storms of secularization (Pinter et al. 2016; Adsera 2006; Kucinski 2005). Since ancient days, for instance, religion has taught about marriage between a man and a woman, and these bore the responsibility of bearing and rearing children (Amien 2014). Nevertheless, numerous forms of relationships and marriages have emerged and are globally practices such that the family unit as most now it is greatly under threat. Even in countries where issues pertaining to LGBTQ are still forbidden and carrying maximum sentences, they are still practiced, though in secrecy, with spirited and consistent campaigns demanding their legalization being maintained (Ryan 2018). From the secular perceptive, the state has no business interfering on its citizen private orientations; hence, should allow people to love whomever they like. On the same, Perry (2018) opined that most religious organization have always been against divorces, but in the recent past, due to issues like home-based violence, divorces have been persistent and the religious grip on this matter is seen as discriminatory and insensitive. The same case applies to issues like maternal health, where the modern world advocates for safe maternal health, where women, and/or couples have the right to decide on issues like abortion without consulting religious teachings (Fullerton et al. 2018).

The challenges that religious dogmas pertaining to different issues have faced from the secular world is an indictment on the religion on its role in the society. The rigidity of different religions in matters of national concerns has seen a perceived decline in religious belief, with people becoming more open to modern challenges; leading to the rise of spirituality over religion. Sherameta and Smith (2017) argue that this spiritualization is advanced by modern religions; like Protestantism, which is thought to be more accommodative and resonating with modern beliefs. They argue that those allow modern thinking and practices and in turn has seen issues like economic development and the reformation of the education sector, among others (Andersen et al. 2017). Beaman et al. (2018) explain that religious practices such as the recitation of prayers in national events and production of religious documents like birth certificate to be admitted to public institutions like schools and other state services must no longer be imposed on people. These old practices are being phased out in order to accommodate people from all walks of life regardless of their belief system.

The decline of the religious imprint on society, where most people do not identify with any organized religion, calls for the reevaluation of some principles; in particular in respect to economic practices catering for inclusive growth (Sofeska 2017; Tomson 2017). For instance, in recent decades, cities are reported to experience rapid population growth with no commensurate increase in economic activities (Joss et al. 2019; Pereira et al. 2017). Through the adoption of revised policies, projects can be made to potentially improve economic growth and equally reduce societal pressures. However, the obsession of celebrating religion as public manifestation is still performed, and the practice of erecting expensive constructions in the form of places of worship on prime spaces are still pursued, where unfortunately those buildings are only used a few hours per week (Poon 2008). To encourage a practical approach, the importance of such buildings could be amplified if they had other purposes supporting economically oriented activities. Where this is not possible, the number of such infrastructures should be reduced in favor of other projects that could concretely help in handling social and contemporary problems like the provision of housing, resource optimization, waste management, population increase and urbanization among many others that confront modern cities. In addition, the reevaluation of current religious ethos should also entail emphasizing on the principles of governance and leadership structure so as to allow for proper and sound management of cities and their fabrics. Religion should not be made to dictate how cities should be managed, or insist on who is to be at the helm of leadership. Instead, they should pave way for the contextualization of principles with both space and time with an aim to address pressing societal concerns.

1.6 Conclusion

Even though there is a decline in religious practice in favor of spirituality, the way cities have been designed throughout time is still being pursued today; highlighting religious principles that may not be applicable in today's information and digital age. Fuelled by an unhealthy relationship between religion and politics, this is impacting on both societal and economic urban fabric leading to a decline in liveability levels. In order to promote more inclusive cities, there is a need for the reevaluation of religious ethos and principles in its relationship to the urban form.

References

Adsera A (2006) Religion and changes in family-size norms in developing countries. Rev Relig Res 47(3):271–286

Ahmed D (2017) Religion—state relations. 2nd edn. International Institute for Democracy and Electoral Assistance (IDEA), Stockholm, Sweden

Aldashev G, Platteau J-P (2014) Relgion, culture, and development. Handbook of the Economics of Art and Culture 2:581–631

Allam Z (2019) The city of the living or the dead: on the ethics and morality of land use for graveyards in a rapidly urbanised world. Land Use Policy 87:104037. https://doi.org/10.1016/j.landusepol.2019.104037

Altmann M, Bunta A, Mazimpaka O (2012) Religion & sustainability: the contribution of religious belief in moving society towards sustainability. Masters, Blekinge Institute of Technology, Karlskrona, Sweden

Amien W (2014) The recognition of religious and customary marriages and non-marital domestic partnerships in South Africa. Routledge,

Andersen TB, Bentzen J, Dalgaard C, Sharp P (2017) Pre-reformation roots of the protestant ethic. Econ J 127:1756–1793

Arikan G, Bloom PB-N (2018) Religion and political protest: a cross-country analysis. Saje J 52(2). https://doi.org/10.1177/0010414018774351

Aroney N (2018) The rule of law, religious authority, and oaths of office. J Law Religion State 6(2–3). https://doi.org/10.1163/22124810-00602003

Aspinall G (2018) Here are the countries where it's still really difficult for women to vote. Grazia. https://graziadaily.co.uk/life/real-life/countries-where-women-can-t-vote/. Accessed 9 Mar 2019

Atkinson QD, Bourrat P (2011) Beliefs about God, the afterlife and morality support the role of supernatural policing in human cooperation. Evol Human Behav 32:41–49

Bai Y, Kung JKS (2015) Diffusing knowledge while spreading God's message: protestantism and economic prosperity in China, 1840-1920. J Eur Econom Assoc 13(4):669–698

Beaman LG, Steele C, Pringnitz K (2018) The inclusion of nonreligion in religion and human rights. Social Compass 65(1):43–61. https://doi.org/10.1177/0037768617745480

Boeck FD (2014) The sacred and the city: modernity, religion, and the urban form in Central Africa. In: Boddy J, Lamberk M (eds) A Companion to the anthropology of religion. John Wiley and Sons, Inc.

Boyer P (2003) Religious thought and behaviour as by-products of brain function. Trends Cogn Sci 7(3):119–124. https://doi.org/10.1016/S1364-6613(03)00031-7

Brown PH, Tierney B (2009) Religion and subjective well-being among the elderly in China. J Socio-Econom 38:310–319. https://doi.org/10.1016/j.socec.2008.07.014

Buckingham JE (2012) The relationship between religions and a secular society. Creed Free Relig Hum Rights 9(3)
Carey J (2018) Spiritual, but not religious?: On the nature of spirituality and its relation to religion. Int J Philos Relig 83(3):261–269
Cesari J (2006) Mosque conflicts in European cities: introduction. J Ethnic Migrat Stud 31(6):1015–1024. https://doi.org/10.1080/13691830500282626
China Travel (2018) Chinese Feng Shui. https://www.chinatravel.com/facts/chinese-feng-shui.htm. Accessed 10 March 2019
Clarke M (2011) Development and religion: theology and practice. Elgar, Cheltenham, UK
Coltrane S, Collins R (2001) Sociology of marriage & the family: gender, love, and property, 5th edn. Wadsworth/Thomson Learning, Belmont, CA
Constitution of the Republic Mauritania Islamic (2006)
Cranenburgh KCV, Louche C, Arenas D (2014) From Stewardship to power: religious organizations and their investment potentials
Dallahite DC, Marks LD, Dalton H (2018) Why religion helps and harms families: a conceptual model of a system of dualities at the nexus of faith and family life. J Family Theor Rev 10(1):1–23. https://doi.org/10.1111/jftr.12242
Day K (2017) Urban space and religion in the United States. Oxford Research Encyclopedias. https://doi.org/10.1093/acrefore/9780199340378.013.470
Dean N, Pryce G (2017) Is the housing market blind to religion? A perceived substitutability approach to homophily and social integration. Urban Studies 54(13):3058–3070. https://doi.org/10.1177/0042098016668779
DeHanas DN, Shterin M (2018) Religion and the rise of populism. Religion, State and Society 46(3):177–185. https://doi.org/10.1080/09637494.2018.1502911
Einstein M (2008) Brands of faith: marketing religion in a commercial age. Routledge, Oxon, OX
Fakhimzade H, Beyki S, Rezaee ED, Shabani M (2016) The comparative study of the intellectual, behavioral, and objective scopes in the "Muslims' City" with the principles of the "Islamic City". Int J Humalt Cult Stud 1:523–541
Fenelon A, Danielsen S (2016) Leaving my religion: Understanding the relationship between religious disaffiliation, health, and well-being. Soc Sci Res 57(49–62). https://doi.org/10.1016/j.ssresearch.2016.01.007
Fernández-Götz M (2018) Urbanization in iron age Europe: trajectories, patterns, and social dynamics. J Archaeol Res 26(2):117–162
Fullerton J, Butler MM, Aman C, Reid T, Dowler M (2018) Abortion- related care and the role of the midwife: a global perspective. Int J Women's Health 10:751–762. https://doi.org/10.2147/IJWH.S178601
Goh DPS, van der Veer P (2016) Introduction: the sacred and the urban in Asia. international sociology 31(4):367–374. https://doi.org/10.1177/0268580916643088
Hage J, Posner BZ (2015) Religion, religiosity, and leadership practices: An examination in the Lebanese workplace. Leadership Organizat Develop J 36(4):396–412. https://doi.org/10.1108/LODJ-07-2013-0096
Harding SR, Flannelly KJ, Weaver AJ, Costa KG (2005) The influence of religion on death anxiety and death acceptance. Mental Health Religion Culture 8(4):253–262
Haynes J (2017) Religion in foreign policy. Oxford Res Encycl. https://doi.org/10.1093/acrefore/9780190228637.013.380
Hirschman C (2007) The impact of immigration on American society: looking backward to the future. institute for human sciences. http://www.iwm.at/transit-online/the-impact-of-immigration-on-american-society/. Accessed 10 Mar 2019
Janssen J (2016) Religiously inspired urbanism: Catholicism and the planning of the southern Dutch provincial cities of Eindhoven and Roermond C. 1900 to 1960. Urban History 43(1):135–157. https://doi.org/10.1017/S0963926815000139
Joo CG (2015) Marriage and sexuality in terms of Christian theological education. Soc Behavior Sci 174:3940–3947

Joss S, Sengers F, Schraven D, Caprotti F, Dayot Y (2019) The smart city as global discourse: storylines and critical junctures across 27 cities. J Urban Technol 26(1):3–34. https://doi.org/10.1080/10630732.2018.1558387
Kang X (2009) Two temples, three religions, and a tourist attraction: contesting sacred space on China's ethnic frontier. Modern China 35(3). https://doi.org/10.1177/0097700408329600
Kauffman E (2010) Shall the religious inherit the earth?. Profile, London
Kirchmaier I, Prüfer J, Trautmann ST (2018) Religion, moral attitudes and economic behavior. J Econ Behav Organ 148:282–300. https://doi.org/10.1016/j.jebo.2018.02.022
Kislali H, Kavaratzis M, Saren M (2016) Rethinking destination image formation. Int J Cult, Tour Hosp Res 10(1):70–80
Koehrsen J (2017) Conceptualizing roles of religion in sustainability transitions. CiRRuS Working Paper No. 9. Universität Bielefeld,
Koenig HG (2012) Religion, spirituality, and health: the research and clinical implications. ISRN Psychiatr: 1–33. https://dx.doi.org/10.5402/2012/278730
Kosher H, Ben-Arieh A (2017) Religion and subjective well-being among children: a comparison of six religion groups. Child Youth Serv Rev 80:63–77. https://doi.org/10.1016/j.childyouth.2017.06.049
Kucinski MA (2005) New York's recognition of same-sex marriages. Family Law Quarter 39(3):841–858
Kumar P (2017) Participation of women in politics: worldwide experience. J Human Soc Sci 22(12):77–88
Kumar V, Aruna R (2018) A study on Sthalavrikshas in Temples of Madurai District, Tamil Nadu. J Sci Res Publ 8(7):295–301
Laeheem K (2018) Relationships between Islamic ethical behavior and Islamic factors among Muslim youths in the three southern order provinces of Thailand. Kasetsart J Soc Sci 39(2):305–311. https://doi.org/10.1016/j.kjss.2018.03.005
Lakpura (2018) Buddhist Sites of Sri Lanka. Lanka. https://lanka.com/about/attractions/buddhist-sites/. Accessed 10 Mar 2019
Lehrer EL (2004) Religion as a determinant of economic and demographic behavior in the United States. Popul Dev Rev 30(4):707–726
Leigh I, Hambler A (2014) Religious symbols, conscience, and the rights of others. Oxford J Law Religion 3(1):2–24. https://doi.org/10.1093/ojlr/rwt048
Liu W (2018) Protection of religious signs under trademark law: a perspective of China's practice. Religions 8:246–259. https://doi.org/10.3390/rel8110246
Llavador H (2017) Suffrage rights. Oxford Res Encycl. https://doi.org/10.1093/acrefore/9780190228637.013.5
Lynch K (1981) A theory of good city form. The MIT Press, Cambridge
Mark JJ (2009) Ancient Rome. Ancient History Encyclopedia. https://www.ancient.eu/Rome/. Accessed 10 Mar 2019
Markham I (2012) Religion and ethics. Encycl Appl Ethics. 748–756. https://doi.org/10.1016/B978-0-12-373932-2.00235-0
Mazumdar S, Mazumdar S (2013) Planning, design and religion: America's changing urban landscape. J Arch Plan Res 30(3):221–243
McCleary RM, Barro RJ (2006) Religion and political economy in an international panel. J Sci Study Relig 45(2):149–175
Mercier B, Kramer SR, Shariff AF (2018) Belief in God: why people believe, and why they don't. Curr Direct Psychol Sci 27(4). https://doi.org/10.1177/0963721418754491
Mosse D (2018) Caste and development: Contemporary perspectives on a structure of discrimination and advantage. World Dev 110:422–436. https://doi.org/10.1016/j.worlddev.2018.06.003
Mumford L (1961) The city in history. Brace and World, New York
Mylek I, Nel P (2010) Religion and relief: the role of religion in mobilizing civil society against global poverty. New Zealand J Social Sci Online 5(2):81–97. https://doi.org/10.1080/1177083X.2010.519391

Narayanan Y (2013) Religion and sustainable development: analysing the connections. Sustain Dev 21:131–139. https://doi.org/10.1002/sd.1557
Nassar DM, Elsayed HG (2017) From informal settlements to sustainable communities. Alexandria Eng J (in Press). https://doi.org/10.1016/j.aej.2017.09.004
Nath S (2015) Religion and Its Role in Society. IOSR Journal of Humanities and Social Science (IOSR-JHSS) 20 (11):82–85
Nieuwenhuis AJ (2012) State and religion, a multidimensional relationship: Some comparative law remarks. Int J Const Law 10(1):153–174. https://doi.org/10.1093/icon/mos001
Oboh GE (2017) Influence of ethnicity and religion in Nigerian elections and the imperative of media intervention. Sociol Int J 1(3):79–84
Olsen N (2009) Marriage and divorce in Islamic and mormon polygamy: a legal comparison. Intermt West J Relig Stud 1(1):87–106
Orgad L (2010) The preamble in constitutional interpretation. Int J Constitution Law 8(1):714–738. https://doi.org/10.1093/icon/mor010
Ottuh POO, Ottuh JA, Aitufe VO (2014) Distributive justice and inclusive society: role of religion in the equitable distribution of wealth. Eur Scientif J 10(23):196–213
Overstreet DV (2010) Spiritual vs. Religious: perspectives from today's undergraduate Catholics. J Catholic Educat 14(2):238–263. http://dx.doi.org/10.15365/joce.1402062013
Pedersen MH, Rytter M (2017) Rituals of migration: an introduction. J Ethnic Migrat Stud 44(16):2603–2616
Pereira GV, Cunha MA, Lampoltshammer TJ, Parycek P, Testa MG (2017) Increasing collaboration and participation in smart city governance: a cross-case analysis of smart city initiatives. J Informat Technol Develop 23(3):526–553. https://doi.org/10.1080/02681102.2017.1353946
Perry SL (2018) Their fault, not mine: religious commitment, theological conservatism, and Americans' retrospective reasons for divorce. Religions 9(238):1–18. http://dx.doi.org/10.3390/rel9080238
Pew Research Center (2015) U.S. Public Becoming Less Religious. Pew. http://www.pewforum.org/2015/11/03/u-s-public-becoming-less-religious/. Accessed 10 Mar 2019
Pew Research Centre (2015) America's changing religious landscape. http://www.pewforum.org/2015/05/12/chapter-3-demographic-profiles-of-religious-groups/. Accessed 10 Oct 2018
Pinter B, Hakim M, Seidman DS, Kubba A, Di Carlo C (2016) Religion and family planning. Eur J Contracept Reproduct Health Care 21(6):486–495
Poon S-W (2008) Religion, modernity, and urban space: the city god temple in republic Guangzhou. Modern China 34(2):247–275
Powell R, Clarke S (2012) Religion as an evolutionary byproduct: a critique of the standard model. Br J Philos Sci 63(3):457–486. https://doi.org/10.2307/23253409
ProCon (2008) References to "God" in state constitutions. ProCon.Org. https://undergod.procon.org/view.resource.php?resourceID=000081. Accessed 10 March 2019
Proppe SJ (2015) To do with urban planning? https://www.strongtowns.org/journal/2015/11/16/what-do-churches-have-to-do-with. Accessed 6 Oct 2018
Quashigah K (2014) Religion and the republican state in Africa: the need for a distanced relationship. African Human Rights Law J 14(1):78–92
Rakodi C (2012) Religion and development: subjecting religious perceptions and organisations to scrutiny. Develop Practice 22(5/6):621–633
Rashid AG (2018) Religious tourism—a review of the literature. J Hospital Tourism Insights 1(2):150–167. https://doi.org/10.1108/JHTI-10-2017-0007
Reese TJ (1998) Inside the vatican. Harvard University Press, London, England
Rieger J (2013) The ethics of wealth in a world of economic inequality: a Christian perspective in a Buddhist Christian Dialogue. Buddhist Christ Stud 33(3):153–162
Rinallo D, Oliver MA (2019) The marketing and consumption of spirituality and religion. J Manag Spirit Rel 16(1):1–5. https://doi.org/10.1080/14766086.2019.1555885
Ryan JM (2018) Gender identity laws: the legal status of global sex/gender identity recognition. LGBTQ Policy J 8(1–100)

Sankaran S, Sekerdej M, von Hecker U (2017) The role of indian caste identity and caste inconsistent norms on status representation. Front Psychol 8. https://doi.org/10.3389/fpsyg.2017.00487
Seetharaman G (2017) Here's a list of countries where same-sex marriages are legal and illegal. The Economic Times. https://economictimes.indiatimes.com/industry/miscellaneous/heres-list-of-countries-where-same-sex-marriages-are-legal-and-illegal/articleshow/61165415.cms. Accessed 9 Mar 2019
Shadabi L (2013) The impact of religion on corruption. J. Bus. Inq. 12:102–117
Sherameta R, Smith V (2017) The impact of the reformation on the economic development of Western Europe
Shokoohy M (1983) The Sasanian caravanserai of Dayr-i gachīn, South of Ray, Iran. Bull Sch Orient Afr Stud Univ Lond 46(3):445–461
Sofeska E (2017) Understanding the livability in a city through smart solutions and urban planning toward developing sustainable livable future of the city of Skopje. Procedia Environ Sci 37:442–453
Spikard JV (2013) Religion and the state: a comparative sociology. J Contemporary Religion 28(3):530–531
Sullivan WF (1994) Paying the words extra: religious discourse in the supreme court of the United States. Cambridge
Terzidou M, Scarles C, Saunders MNK (2018) The complexities of religious tourism motivations: sacred places, vows and visions. Ann Tourism Res 70:54–65. https://doi.org/10.1016/j.annals.2018.02.011
Theodorou AE (2014) In 30 countries, heads of state must belong to a certain religion. Fact Tank. http://www.pewresearch.org/fact-tank/2014/07/22/in-30-countries-heads-of-state-must-belong-to-a-certain-religion/. Accessed 10 Mar 2019
Tomalin E (2018) Religions, poverty reduction and global development institutions. Palgrave Commun 4(132):1–12. https://doi.org/10.1057/s41599-018-0167-8
Tomson T (2017) Understanding the contextual development of smart city initiatives: a pragmatist methodology. She Ji J Design Econom Innovat 3(3):210–228. https://doi.org/10.1016/j.sheji.2017.11.004
Trimble M (2019) The most religious countries, ranked by perception. U.S. News. https://www.usnews.com/news/best-countries/articles/10-most-religious-countries-ranked-by-perception. Accessed 8 Mar 2019
UN Women (2018) Facts and figures: Leadership and political participation - Women in parliaments. UN Women. http://www.unwomen.org/en/what-we-do/leadership-and-political-participation/facts-and-figures. Accessed 9 Mar 2019
UNESCO Caravanserais: cross-roads of commerce and culture along the Silk Roads. Unesco. https://en.unesco.org/silkroad/content/caravanserais-cross-roads-commerce-and-culture-along-silk-roads. Accessed 10 Mar 2010
Weller PG (2015) Religion and state in marriage, cohoabitation and civil partnership: examples, typologies and contestations from the United Kingdom. In: Brunn S (ed) The changing world religion map. Springer, Dordrecht
Zhang Y, Wei T (2017) Typology of religious spaces in the urban historical area of Lhasa, Tibet. Front Architect Res 6(3):384–400. https://doi.org/10.1016/j.foar.2017.05.001
Zuhaira MAM, Ye-zhuang T (2017) The effect of relious beliefs, participation and values on corruption: survey evidence from Iraq. Int J Adv Comput Sci Appl 8(8):302–305

Chapter 2
Religious Matrimony, Urban Sprawl and Urban Morphology

Abstract With an unprecedented increase in urbanization, urban sprawl is seen as arguably the single biggest threat to the biodiversity of cities and regions by supporting an increasingly high dependency on fossil fuel consumption. However, while the topic of urban sprawl has been widely documented through sciences such as urban planning, ecology, environment, infrastructure, and others, there have been little, to no literature on the correlation between urban morphology and the sociological and religious viewpoints of individual housing units; making the grounds for sprawl. This perspective chapter puts forth the argument that religious matrimony contributes largely to Urban Sprawl through the contemporary dogma of monogamy and nuclear family as being basic social units. In this essence, it highlights unsustainable a fundamental land-use policy fuelled largely by religious practices. While the author does not dispute religious claims, the chapter supports that the sustainability of cities need to be reviewed and questioned through equal terms in religious belief systems; which impact on both urban morphology and society. This chapter is aimed towards urban theorists, anthropologists, and sociologists.

Keywords Urban sprawl · Religion · Matrimony · Sustainability · Society · Urban morphology

2.1 Introduction

While the aftermaths of Urban Sprawl has been widely documented, Bhatta (2010) shares that there are diverging opinions regarding the subject. One commonality is the characteristic of low-density developments beyond the boundaries of cities or urban centers (Habibi and Asadi 2011). Herzog (2015) views urban sprawl as the migration of people from populated urban centers to the suburbs areas; hence, warranting development of low-density residential areas, uncontrolled shopping centers, and curvilinear street patterns. Pendall (1999), Agrawal (2010) and Sudhira and Ramachandra (2007), share a similar viewpoint where they see urban sprawl as unplanned, uncontrolled and unauthorized developments that emerges at the periphery of cities; hence, hindering planned and mixed-use of available spaces in suburban areas. There is a similar variation in defining characteristics of urban sprawl just as in the causing

Z. Allam, *Theology and Urban Sustainability*, SpringerBriefs in Geography,
https://doi.org/10.1007/978-3-030-29673-5_2

Fig. 2.1 Suburb in Le Bausset, France. Photo by Raphaël Biscaldi

factors. It has been noted that land prices are relatively high in urban centers, causing unaffordability coupled with the lack of property development; resulting in a low demand and low supply. With an increasing housing need driven by affordable pricing demands, suburban areas flourish encouraging a brand of social and inclusive lifestyle where each house will have its own parking, garden and private spaces. The suburbs of Le Bausset in France represent this perfectly (Fig. 2.1).

This lifestyle has been often branded as the suburban dream lifestyle and promoted by star architects and urbanists supporting automobile dependence (Salingaros 2000). Even though the need for automobile demands substantive infrastructure, those are often viewed as public infrastructure and thus offset from development costs from private sector development (Allam and Newman 2018). As such, suburban life is rendered cheaper than urban life. This is further supported by the fact that the urban life is relatively expensive, and families cannot accommodate for a nuclear family structure, as there are is a notable lack of amenities in the form of open spaces. Moreover, this need for individual housing, in the form of nuclear housing structures, is hailed by societal dogmas in the form of civil union and religious practices in the form of religious matrimony. This need for housing, coupled with urban economics, encourages migration to suburbs and urban sprawl; hence posing threats to biodiversity and fragile and unique ecosystems. This opinion chapter explores this phenomenon and further ponders on the ethics and philosophy of such a practice.

2.2 A View on Family

The family constitution has considerably evolved in the last thousands of years (Okasha et al. 2012). A review of anthropological family structures showcases that most societies support a patriarch nuclear family, which did not limit to relatives alone. This can be seen in the Roman society for instance, where a family consisted of a father, his wife(s) and children and sometimes, other relatives. In the case where the family was well endowed, the family setup also included slaves and servants (Rawson 2011). In ancient Greece, and what was then feudal Europe, individuals who contributed to the financial and operational structure of a household was regarded as a member of that family (Coltrane and Collins 2001). This meant that even servants and slaves were family members. One characteristic of this type of family structure is that the head of the household, the authoritative figure, could marry as many wives as he wished, and children born in his household, even by the servants and slaves, belonged to him. This meant that everyone who was part of the household lived in the same compound where the families encouraged shared amenities. A similar structure is seen in African setups where polygamous families, where one husband married two or more wives, were seen to live in the same compound, which consisted of a number of units where some dedicated to housing wives, children and servants (Mair 1953).

For Coontz (2005), ancient families were not borne out of love between parties but, out of necessity; political, social and/or economic needs. This highlights how wealth was created and why household heads could financially cater for numerous wives and children, which was synonymous with an upper-class structure (Allen 2009). For political or economic gains, a given family could exchange their children or, more specifically, their girls, to another to strengthen their political ties or sometimes, to cease political tensions.

Today, family structures have significantly evolved and a family is considered as the smallest unit of the society (Forsyth 2014), usually starting by the marriage between two people, which can be between people of the opposite sex, though other forms are being recognized, there is extreme resistance; especially when it comes to marriage between people of the same sex (Kucinski 2005; Bennett 2017; Gates 2015; Brown et al. 2014; Joo 2015). The modern definition classifies single-parent households (Baker 1999) and even single-person households (Bachman and Barua 2015). Coontz (2005), argues that this change results from evolution from the eighteenth century family setup. However, one dimension that is constant in its view of a family is that of religion. It is worth noting that despite the nature or form of marriage, though some may seem controversial, marriages are influenced by religious beliefs; hence, most families adopt religious teachings especially in regard to housing, environment, economy, and politics among other things, and religious practices can, over time, be made to have a substantial impact on land use policy as showcased later in this chapter.

2.3 A Review of Religious Matrimony

One religious practice that has had an impact on urbanism is that of marriage. The teachings of different religions perceive marriage in the same way: that between a man and a woman. In the Christian context, this conviction is drawn from the bible, where it is stated that God created a man and a woman and commanded them to be fruitful ("Genesis 1: 27–28"). From this scripture, the understanding is that God established a family unit similar to the modern nuclear family. Christian beliefs, established by the teaching of Jesus and Paul, encouraged marriage between a man and only one wife. The church advances this by being a witness of vows between a man and a woman united together in marriage (Joo 2015). This is nicely depicted by the photograph of Josh Applegate in Fig. 2.2. Despite this, biblical figures like Abraham (Genesis 16), David, Solomon, and Jacob among others had many wives and even concubines (Breckenridge 2004; Robert 1967).

Just like in Christianity, Islam advocates for a marriage between a man and a wife. However, polygamy is permissible, but monogamy is the most predominant. According to scholars, as noted by Olsen (2009), the primary goal of marriage as per Allah's will is for the married to fill the earth will faithful Muslims. Polygamous was and is still viewed by many Muslim scholars as a way to ensure that the Muslim community has fewer widows and orphans (Esposito 2001). The Mormons, like their religious counterparts, also supports monogamous marriages. However, members of that community cited psychological, financial and emotional burdens as the reason not to marry many wives (Olsen 2009). Hinduism, according to Carroll (1997), ruled out polygamy in 1955. Judaism also rejected the concept of plural marriages in the early Middle Ages and instead, advocated for monogamous family, with the family tasked with the responsibility of siring responsible children (Goldfeder 2013). Catholicism also only recognizes a marriage that involves a man and his wife and emphasis on children.

The religious clarion call is for monogamous marriages and a family is seen as an establishment whose origin is the union between a man and woman. Even though various religious groups have differing views on family planning, the undoubtful expectation is the need to bear as many children as one can manage (Pinter et al. 2016).

The possibility of bearing as many children as possible attracts a number of issues. A key one is the need for decent housing, better health care, and quality education. This also means an increase in population; hence, migration, need for housing, schools, traffic congestion, demand for food and health care, among many other factors. With the nature of the urban centers and the costs of living in these areas, these families are forced to seek alternative places to settle. Suburbs are seen as the most attractive since they provide ample spaces for larger family homes, lawns and fresh air away from the city life (Hackworth and Stein 2012). The most unfortunate part is that most suburbs are often outside planned perimeters of urban areas; hence, provisions of services such as health care, street lighting, waste management, security, water distribution, and controlled planning becomes challenging (Agrawal 2009),

Fig. 2.2 Wedding ceremony depicting ceremonial vow to god over one another. Photo by Josh Applegate

thus, qualifying as being classified as urban sprawl. The trend of families to migrate to suburbs, either from organized cities or even from the rural areas impacts negatively on agricultural farms by altering the land use and ultimately affecting the food policy and environmental management (Guiling et al. 2009). The uncontrolled nature of most suburban areas means that they extend far and beyond, and as this happens, there is increased conversion of available spaces to residential areas.

2.4 Religious Matrimony and the Single Housing Units

Religion has also been at the forefront of issues pertaining to family and child-bearing since time immemorial. A key issue is the advocacy by different religious groups on the size of each household. In 1995, Pearce et al. (2015) embarked on a task that involved explaining how religious groups differ in their call for family sizes that their followers should adhere to. According to them, the religiosity of individuals and families are key factors in determining the preferred family sizes. They identified three factors that explain the fertility rate and birth rates among different religious group. These include contraceptive use, minority groups trends of delaying or forgoing births and the inequality in socioeconomic between groups. Adsera (2006), supports that different religious groups stress on families differently; hence, the variation in the sizes of families as per their religious affiliation. From an analysis of 16,000 adults drawn from 13 developed countries. The researcher highlights that Conservative Protestants and Catholics affiliated families were relatively larger compared to Mainline Protestants and were extremely small among families with no religious affiliation. The findings showcase that the said religious groups advocated for marriages between a man and a woman and the need for them to have children according to how God would bless them. Similar results were reported in a research conducted in America by Pew Research Centre (2015).

In the two cases (Adsera 2006; Pew Research Centre 2015), it is clear that difference in religious beliefs and teachings can impact on various demographic groups. This is increasingly important as it relates to population growth from a religious viewpoint. It has been reported that the Mainline Protestants, tend to delay marriage and childbirth; hence, the size of families associated with them are relatively small. On the contrary, those affiliated with Catholics, Mormon, and Conservative Protestants, Jewish and Muslim among other traditional religions tend to have larger families because these religions discourage the use of contraceptives and encourages to have many children (Adsera 2006; Pew Research Centre 2015; Pearce et al. 2015; Spenser 2010; Renzi 1975). Buddhism, like those unaffiliated with any religious groups, are reported to have smaller families since they do not consider child-bearing a duty (Bista 1991).

With the changing economic landscape, and the increase in psychological and emotional burdens and the increased campaigns on the need to have fewer children, families across the religious divides are becoming smaller (McQuillan 2004). Olsen (2009) shares this is even felt among the Mormons, who were free to marry as many as they wish and encouraged to have many children. Religious beliefs are seen to adapt to local challenges. Catholicism, for instance, loosened their stand on contraceptive use and the need to marry when one is young. The argument by Olsen (2009) captures the current religious evolutionary trend. Moreover, unlike in the past where families lived in large household with a leader, mainly the father, families are now living in single-unit houses comprising of parents and their children (Kumar and Tiwari 2016).

Literature further supports the omnipresent role of religion and society. Religiosity plays such a role that people have been known to establish settlements around places of worship, and today this is reflected by the fact that places of worship need to be easily accessible (McCoy et al. 2016). Moreover, those are places where society is being taught to save, spend wisely, invest in family (own properties) to be able to live decent lives as prescribed in religious books (Dilmaghani 2017a). This led to the disintegration of families into nuclear units and the emergence of single-family financial dependence (Lehrer 2004; Keister 2008). The emphasis on education is a key proponent and driver of the increased income (Chiswick and Huang 2008; Dilmaghani 2017b). The emphasis on property ownership means that each household feels obliged to secure a space of their own. This leads to the contemporary notion of high prices and low availability of land in urban areas; hence, suburban life becomes the most potent alternative for many families.

2.5 The Place of Worship and Urban Sprawl

Jacobsen (2006), stated that, traditionally, churches were built at the heart of neighborhoods to serve the communities therein. This later evolved to people relocating, or building around churches (Orsi 1999). This can be seen in European old cities, where examples are Hvar in Croatia (Fig. 2.3) and Bern in Switzerland (Fig. 2.4).

According to Agrawal (2009), religious groups, just like individuals are experiencing the same challenges urban dwellers face. High costs of land, limited spaces, high construction costs, congestion and growth (Park 2004). These factors, among others, encourage religious groups to seek land far from city centers to build structures that resonate with their need to "serve" an increasing population away from urban centers. This is further supported by Cobbinah and Korah (2014) and Bouma and Hughes (2000), as areas outside the established cities are considered due to space availability, cheaper price of land and proximity to basic urban facilities (Shehu and Molyneux-Hodgson 2014). This then encourages faithful practitioners, with an aim of being near to their place of worship, to buy and construct homes near their religious establishments, and before long, a new neighborhood emerges (Ley 2008).

Herzog (2015) noted that the suburbs emerge further from the city centers through urban sprawl and are mostly accessible through roads, which encourages excessive use of automobile. While it was noted that some parishioners erect homes to be nearer to churches, there were also cases where temporary camps or unplanned neighborhoods emerge in view of increasing presence and contribution to religious

Fig. 2.3 Hvar, Croatia. Photo by Marcus Löfvenberg

meetings (Kinney and Winter 2006). This relationship between places and worship and regular meetings is seen, for example, in Islam and Christianity. In Islam, there are five scheduled prayers (Salaat) essential to Muslim faithful and access to a mosque is viewed as a necessity rather than a luxury (Nason 2006). Ebaugh and Chafetz (2000) further support that practicing Catholics would rather be near a church to attend scheduled masses. Interestingly, the case of urban sprawl can be seen from the perspective of graveyards too as those increase in size and fight the dynamics of urban sprawl (Allam 2019). One could build arguments that both are driven by religious principles. This is explored further in Chap. 3.

Proppe (2015) confronts the teachings of religious groups in regard to neighborhoods and how the suburban locations of such religious group are contradictory. According to her, faith teaches to love their neighbors and care for the environment. Nevertheless, the location of churches means that attendees have to travel from far; hence, do not get time to neither know their neighbors nor care for their environment. Similarly, the same churches are forced to build mega parking lots to accommodate the many vehicles that their faithful use, which can be increasing counterproductive of space usage, as shown in Fig. 2.5 These trends of conspicuous consumption of resources, especially land and biodiversity, and the tendency to enjoy privacy are synonymous with urban sprawl and, as Proppe (2015) noted, is contradictory to the teaching of the very religious groups that advocate for care and love.

Fig. 2.4 Bern, Switzerland. Photo by Joel and Jasmin Førestbird

Fig. 2.5 Parking lot in New Jersey, United States. Photo by Omer Rana

2.6 Religion, Society, and Sustainability

Religion is seen as a complementary figure in the provision of services such as health care, education, social welfares, security and environmental conservation (Palmer and Finlay 2003; Rakodi 2012). However, religion has also been argued to be a hindrance to the adoption of the Sustainable Development Goals (SDGs), as prescribed by the United Nations. The stance of religion in regards to issues like discrimination of women, support violence (Ranstorp 1996), contribute to civil strife and support of human rights abuse (Kauffman 2010) has long been criticized. Nevertheless, in the recent past, there have been tremendous changes even some extreme religious stands such as discrimination of women are weathering down. Women are now allowed to drive in the Kingdom of Saudi Arabia (BBC News 2018), they are allowed to hold political positions (Tremblay 2007) and many religious groups are in the forefront in fighting terrorism (Badey 2002).

In regard to environmental conservation, it has been seen that religious groups advocate for proper and optimal use of resources (Tomalin 2012). Hinduism considers nature as the body of god and worships animals; birds and sea animals as different gods and goddess; hence, protecting the biodiversity. Followers of Hinduism are under strict moral guidelines regarding how to utilize the environment. Islam teaches that man should consider the balance and integrity of the universe by being the custodian of the environment and should not draw from it more than required. According to Islam, everything in the universe was created for the glory of Allah; hence, a man should not participate in the destruction of godly creations.

Christianity and Judaism also teach that God owns everything and man was created to take care of God's creation as recorded by. Many religious beliefs teach the need for restraint in using the resources (Clarke 2011). Palmer and Finlay (2003), contends that religion plays a key role in imparting and stressing the need for ethical and moral living. According to them, religious practitioners are compelled to undertake conservation as their way of life; hence, a key proponent of sustainability. Their arguments are akin to those of Koehrsen (2017) who posited that, in ecology, religion either directly plays a key role in sustainability. He also argues that other agents of sustainability, sometimes, indirectly uses religion to communicate their agenda in what he perceived as "green religion". Altmann et al. (2012), argues that religious groups are able to achieve sustainable drives since they have successfully managed to instil behavioral change among their followers. This makes them superior agents in advocating for sustainable development as expounded by Palmer and Finlay (2003). Narayanan (2013) pushes the debate of religion and development further by arguing that religious contributions in sustainable development should be viewed both in terms of its constructive and destructive potential. He argues that, through its values, its social and ecological activism and its impact on individual's self-development, religion should not be overlooked in the discussion about development, planning and policy formulation.

In analyzing the reality of the current developmental trends such as urban sprawl, the reality of religious principles discussed above are put to test. One such principle is that expounded by Koehrsen (2017) and Mcleod and Palmer (2015), who view religion as an agent of self-development. Their arguments are valid since most religions advocate for human dignity; pointed by factors like having a decent shelter, access to health care and affordable education, safe environment and access to social amenities among others. On the other side, in a bid to actualize religious teachings, conflicts arise with sustainability principles. For instance, to secure a decent shelter for each family, more housing units are required and unfortunately, with the dwindling spaces and exorbitant prices in cities, suburbs and other spaces outside the city stands as the obvious and natural choice.

As discussed above, designated areas for human settlements do not have sufficient spaces to accommodate everyone (Nassar and Elsayed 2017; Allam 2018; Allam and Jones 2018). Also, the few available spaces are expensive and within areas that are already congested and full of city life, which, according to religious teaching contradicts their values. Therefore, lands bordering towns and cities become the potential areas where aspirations of having decent housing could be actualized. This happens without caring for the purpose for which such parcels of land were conserved; hence, the principle of sustainable development is obscured (Dovey 2015).

Another important factor that obscures the religious principles is the disharmony between different religious groups (Dean and Pryce 2017). This disharmony has resulted in civil strives, poverty, human displacement, destruction, terrorism and distrusts. In addition, these disharmony means that the faithful of different religious groups require a neighborhood of their own, and this is detrimental in land use policy since most are skewed toward urban sprawl which is not easy to control. In the presence of these, sustainable development cannot be achieved. To overcome this, unconditional collaborations between different religious groups, governments, and agencies need to be seamless (Hewstone and Hughes 2015). Otherwise, the individual initiatives that each religious group take toward sustainable development will not be successful.

2.7 Conclusion

This chapter offers a new perspective through the exploration of the influence of religion on matrimony and its link to urban sprawl. It is seen that nuclear families are a product of religious trends; which has been enforced as a standard in social practice. This has encouraged the mushrooming of single-unit housing which contributed to urban sprawl and to fossil fuel consumption and an unhealthy lifestyle. While this chapter does not dispute religious practices, it calls for the reevaluation of urban sprawl through numerous facets; including social and religious grounds. The sustainability of cities is shown to be intricately linked to behavioral patterns, which may be infused by religious beliefs.

References

Adsera A (2006) Religion and changes in family-size norms in developing countries. Rev Religious Res 47(3):271–286

Agrawal S (2009) New ethnic places of worship and planning challenges. Plan Canada

Agrawal S (2010) Neighbourhood patterns and housing choices of immigrants. Region of Peel, Ontario, Canada

Allam MZ (2018) Redefining the smart city: culture. Case Study of Port Louis, Mauritius. Curtin University, Perth, Australia, Metabolism and Governance

Allam Z (2019) The city of the living or the dead: on the ethics and morality of land use for graveyards in a rapidly urbanised world. Land Use Policy 87:104037. https://doi.org/10.1016/j.landusepol.2019.104037

Allam Z, Jones D (2018) Promoting resilience, liveability and sustainability through landscape architectural design: a conceptual framework for Port Louis, Mauritius; a Small Island Developing State. Paper presented at the IFLA World Congress Singapore 2018, Singapore

Allam Z, Newman P (2018) Economically incentivizing smart urban regeneration. Case Study of Port Louis, Mauritius. Smart Cities 1(1):53–74

Allen TD (2009) The Ancient Egyptian family: kinship and social structure. NY, Taylor & Francis Group, New York

Altmann M, Bunta A, Mazimpaka O (2012) Religion & sustainability: the contribution of religious belief in moving society towards sustainability. Masters, Blekinge Institute of Technology, Karlskrona, Sweden

Bachman D, Barua A (2015) Single-person households: Another look at the changing American family. Deloitte Insights

Badey TJ (2002) The role of religion in international terrorism. Soc Facus 35(1):81–86

Baker DG (1999) the increase of single parent families: an examination of causes. Policy Sci 32(2):175–188

BBC News (2018) Saudi Arabia issues first driving licences to women. https://www.bbc.com/news/world-middle-east-44367981. Accessed 9 Oct 2018
Bennett NG (2017) A reflection on the changing dynamics of union formation and dissolution. Demograph Res 36:371–390
Bhatta B (2010) Analysis of urban growth and sprawl from remote sensing data, vol 1. Springer, Berlin Heidelberg
Bista DB (1991) Fatalism and Development in Nepal. Orient Longman ltd, India
Bouma G, Hughes P (2000) Religious residential concentrations in Australia. People Place 8:18–27
Breckenridge JP (2004) Old testament teaching on polygamy. Torch Trinity J 7:10–30
Brown RM, Bristol JC, Clark ES, Pasquier M, Cressler MJ, Selka S (2014) Black catholicism. J Africana Religions 2(2):244–295
Carroll L (1997) Religious conversion and polygamous marriage. J Indian Law Institute 39(4):272–280
Chiswick BR, Huang J (2008) The earnings of American Jewish men: Human capital, denomination, and religiosity. Journal for Scientific Study of Religion 47(4):694–709
Clarke M (2011) Development and religion: theology and practice. Elgar, Cheltenham, UK
Cobbinah PB, Korah IP (2014) Religion gnaws urban planning: the geography of places of worship in Kumasi, Ghana. Int J Urban Sustain Develop 8(2):93–109. https://doi.org/10.1080/19463138.2015.1074581
Coltrane S, Collins R (2001) Sociology of marriage & the family: gender, love, and property, 5th edn. Wadsworth/Thomson Learning, Belmont, CA
Coontz S (2005) Marriage, a history: from obedience to intimacy or how love conquered marriage. Viking Penguin, New York, NY
Dean N, Pryce G (2017) Is the housing market blind to religion? A perceived substitutability approach to homophily and social integration. Urban Studies 54(13):3058–3070. https://doi.org/10.1177/0042098016668779
Dilmaghani M (2017a) Religious identity and real estate wealth accumalation: evidence from Canada. Housing Stud 2–31. https://doi.org/10.1080/02673037.2017.1364714
Dilmaghani M (2017b) Religiosity and labour earnings in Canadian provinces. J Labor Res 38(1):88–99
Dovey K (2015) Sustainable informal settlements? Soc Behavior Sci 179(2015):5–13
Ebaugh HR, Chafetz JS (2000) Religion and the new immigrants: continuities and adaptations in immigrant congregations. Altamira, Walnut Creek, CA
Esposito JL (2001) Women in Muslim family law. Syracuse University Press, Syracuse, NY
Forsyth A (2014) Defining marriage. Sound Interdiscipl J 97(3):297–322
Gates GJ (2015) Marriage and family: LGBT individuals and same-sex couples. Fut Child 25(2):67–87
Goldfeder M (2013) The story of Jewish polygamy. Columbia J Gender Law 26(2):234
Guiling P, Brorsen W, Doye D (2009) Effect of urban proximity on agricultural land values. Land Econom 85:252–264
Habibi S, Asadi N (2011) Causes, results and methods of controlling urban sprawl. Proc Eng 21:133–141)
Hackworth J, Stein K (2012) The collision of Faith and economic development in Toronto's inner suburban industrial districts. Urban Affairs Rev 48(1):37–63. https://doi.org/10.1177/1078087411420374
Herzog LA (2015) Global suburbs: urban sprawl from the Rio Grande to Rio de Janeiro, vol 1. Taylor & Francis, Abingdon, Oxon
Hewstone M, Hughes J (2015) Reconciliation in Northern Ireland: the value of inter-group contact. British J Psychiatr Int 12(3):65–67
Jacobsen E (2006) New urbanism and the Church: an overdue partnership. Stained Glass Urban, 43–49
Joo CG (2015) Marriage and sexuality in terms of Christian theological education. Soc Behavior Sci 174:3940–3947

Kauffman E (2010) Shall the religious inherit the earth?. Profile, London
Keister LA (2008) Conservative protestants and wealth: how religion perpetuates asset poverty. Am J Sociol 11(5):1237–1271
Kinney N, Winter W (2006) Places of worship and neighbourhood stability. J Urban Affairs 28:335–352
Koehrsen J (2017) Conceptualizing roles of religion in sustainability transitions. CiRRuS Working Paper No. 9. Universität Bielefeld
Kucinski MA (2005) New York's recognition of same-sex marriages. Family Law Quarter 39(3):841–858
Kumar M, Tiwari PSN (2016) Structural influence of family on religious orientation, quality of llife and happiness. Int J Appl Psychol 6(4):94–99. https://doi.org/10.5923/j.ijap.20160604.03
Lehrer EL (2004) Religion as a determinant of economic and demographic behavior in the United States. Populat Develop Rev 30(4):707–726
Ley D (2008) The immigrant Church as an urban service hub. Urban Stud 45(10):2057–2074
Mair LP (1953) African Marriage and Social Change. In: Phillips A (ed) Survey of African marriage and family life. Oxford University Press, London, pp 1–177
McCoy J, Kirova A, Knight WA (2016) Gauging social integration among Canadian Muslims: a sense of belonging in an age of anxiety. Can Ethn Stud 48(2):21–52
Mcleod E, Palmer M (2015) Why conservation needs religion. Coast Manag 43(3):238–252. https://doi.org/10.1080/08920753.2015.1030297
McQuillan K (2004) When does religion influence fertility? Populat Develop Rev 30(1):25–56
Narayanan Y (2013) Religion and sustainable development: analysing the connections. Sustain Dev 21:131–139. https://doi.org/10.1002/sd.1557
Nason R (2006) Start-up religion: visiting a Mosque. Evans Brothers Limited, Chiltern Street, London
Nassar DM, Elsayed HG (2017) From informal settlements to sustainable communities. Alexandria Eng J (In Press). https://doi.org/10.1016/j.aej.2017.09.004
Okasha T, Elkholy H, El-Ghamry R (2012) Overview of the family structure in Egypt and its relation to psychiatry. Int Rev Psychiatr 24(2):162–165. https://doi.org/10.3109/09540261.2012.658030
Olsen N (2009) Marriage and divorce in Islamic and mormon polygamy: a legal comparison. Intermountain West J Religious Stud 1(1):87–106
Orsi AR (1999) Gods of the city: religion and the American urban landscape. Indiana University Press, Bloomington
Palmer M, Finlay V (2003) Faith in conservation: new approaches to religions and the environment. World Bank Publications, Washington, D.C.
Park C (2004) Religion and geography. In: Routledge companion to the study of religion. Routledge, London,
Pearce LD, Brauner-Otto S, Ji Y (2015) Explaining religious differentials in family size preferences: Evidence from Nepal in 1996. Population Study 69(1):23–37. https://dx.doi.org/10.1080/00324728.2014.995695
Pendall R (1999) Do land use controls cause sprawl. Environ Plann 26:555–571
Pew Research Centre (2015) America's changing religious landscape. http://www.pewforum.org/2015/05/12/chapter-3-demographic-profiles-of-religious-groups/. Accessed 10 Oct 2018
Pinter B, Hakim M, Seidman DS, Kubba A, Di Carlo C (2016) Religion and family planning. Eur J Contracept Reproduct Health Care 21(6):486–495
Proppe SJ (2015) To do with urban planning? https://www.strongtowns.org/journal/2015/11/16/what-do-churches-have-to-do-with. Accessed 6 Oct 2018
Rakodi C (2012) Religion and development: Subjecting religious perceptions and organisations to scrutiny. Develop Practice 22(5/6):621–633
Ranstorp M (1996) Terrorism in the name of religion. J Int Affairs 50(1):41–62
Rawson B (2011) Families in the Greek and Roman worlds, vol 1st. Blackwell Publishing Ltd, Malden, MA

Renzi M (1975) Ideal family size as an intervening variable between religion and attitudes towards abortion. J Scientif Stud Religion 14:23–27
Robert H (1967) Polygamy and the Bible. Int Rev Mission 56:205–212
Salingaros NA (2000) Complexity and urban coherence. J Urban Design 5(3):291–316. https://doi.org/10.1080/713683969
Shehu M, Molyneux-Hodgson S (2014) Faith communities and environmental degradation in North-east Nigeria. Int J Environ Sustain 10(1):27–40
Spenser TM (2010) The Missouri Mormon experience. University of Missouri Press, Columbia
Sudhira HS, Ramachandra TV (2007) Characterising urban sprawl from remote sensing data and using landscape metrics. In: 10th International conference on computers in urban planning and urban management, Iguassu Falls, PR Brazil, July 11–13, 2007
Tomalin E (2012) Thinking about faith-based organisations in development: where have we got to and what next? Develop Practice 22(5/6):689–703
Tremblay M (2007) Democracy, representation, and women: a comparative analysis. Democratization 14(4):533–553. https://doi.org/10.1080/13510340701398261

Chapter 3
Urban and Graveyard Sprawl: The Unsustainability of Death

Abstract The subject of death can be difficult to deal with due to the varying number of religious practices and their difference in beliefs regarding afterlife. However, many of them converge towards the burying of the dead to help in the ascent towards a peaceful afterlife. This has been a practice led since time immemorial. This practice, however, while it brings positive emotional responses and helps in the way people may find closure with the deceased, can be argued to be unsustainable in the way that it consumes space. This is especially emphasized in the case of increasing population growth and that of urbanization leading to scarcity of land. To build this case, This chapter explores three cemetery cases and, explores through a review of literature, the contemporary discourse on ethics and morality as to the question who gains more importance, the living or the dead pertaining to the questions of increased urbanization and the problems it welcomes regarding land scarcity, gentrification, and housing unaffordability. This chapter further showcases how there are now emerging forces between urban and graveyard sprawl, which can cause social tensions and impact on property value, and hence; discourage residential areas around graveyards, even if those are in highly populated areas.

Keywords Religion · Death · Sustainability · Land use · Graveyards · Cities · Urban sprawl

3.1 Introduction

It is widely seen that most people maintain a close relationship among each other, and this transcends life and manifests in various forms after death. Though not a relationship that can expressly be explained, this relationship is characterized by behaviors, ceremonies and actions observed. According to Johnson et al. (2018), of the many things that can be said to be pointers to this assertion is the practice of burying the dead, which is practiced in many different parts of the world, with most of such practices accentuated by religious faiths. Though some choose to have their loved ones cremated; these too are said to take home ashes of the dead, which is also a sign of this bond that is never completely broken as is expressed by Mathijssen (2017). But, in this chapter, the concern is more on those who choose to bury the

Z. Allam, *Theology and Urban Sustainability*, SpringerBriefs in Geography,
https://doi.org/10.1007/978-3-030-29673-5_3

dead, and the impact of such practices has on the living, especially in respect to the diminishing spaces in urban areas. If we are to look from a historical lens as is described by Beard and Burger (2017), it can be observed that most burials happened in home compounds or delineated plots within private properties or in other cases, in churchyards dedicated within religious places of worship for burial purposes of church members, often, with time, open to the public. As time progressed and land scarcity became more apparent and government policies on disposal of the dead became stringent in most countries and cities, people started relying on public or privately owned cemeteries, but over time, these too have become scarce and opened new kinds of problems (Swensen 2018; Hayward 2018; Tudor et al. 2013).

In the past, cemeteries and graveyards were located on the periphery of cities. But, as demand for burial plots in them started to become limited; expansion of such cemeteries and graveyards became inevitable. Unfortunately, most expansions pointed toward the centres of highly urbanized areas, or at times are found to be in prime real estate spots (Niţă et al. 2014), as seen in Fig. 3.1; showcasing a cemetery enjoying seafront views.

Therefore, these expansions prompted demand for more spaces, and, as is the case in most urban areas, the delimitated zones for urban areas are almost static; hence, any change in land use prompted demand for more land. In this case, the expansion of graveyards in cities leads to an increase in the cost of land. In most cases, as explained by Bennett and Davies (2015), such increases result into high property prices and this, in turn, translates into accelerating gentrification.

In some cases, the gentrification phenomena happens right within the city where the case of cemetery expansion takes place, or in other cases, wealthier people living near such cemeteries are forced to look for living spaces elsewhere, and the most potent areas are those occupied by poor and less affluent, who are willing to sell out. When properties are sold, Hochstenback and Musterd (2018) explain that those undergo extensive renovations which, with time, changes the dimension of the surrounding areas. Those are catalysts in attracting more affluent households; hence, affecting the cost of living in selected areas, which can be problematic as, Chapple (2016) supports that gentrification negatively impacts on the livelihood of residents. Similarly, it is worth to note that graveyards and cemeteries unfairly occupy large tracts of lands in the urban areas, exacerbating problems already worsened by population increase and urbanization. It also raises the questions of whether it is ethically and morally right for them to occupy such larges spaces, while the rest of the urban fabric is seen to continue constricting. On this front, Gade (2015) explains that such questions can only be answered from a religious perspective since the idea of burying the dead for most people is based on religious teachings; relating to philosophies about life and the afterlife.

In the previous chapters, it was clearly noted that most religious faiths are of the opinion that the dead only transitions into the next life and after a believed upcoming apocalypse, they will rise again (Shepkaru 1999). This gave rise to ever-expanding graveyard sites, with varying practices. On the other hand, share that the concept of cremation is seen as part of the answer to reducing the pace at which demand for burial spaces are confronting the urban areas is not embraced. However, Rotar and

Fig. 3.1 Waverley Cemetery, Australia. *Source* Connor Meakins

Todor (2013) and also Geisler and Potter (1998) highlight that cremating a body is seen in some quarters as interfering with God's creation and perceived to attract his wrath on those who are living. Another avenue that could ease the ethical and moral questions of having cemeteries occupying large spaces at the expense of the living is the reuse of burial plots at regular intervals.

While this is noted, it is seen as a rare occurrence. Rugg and Holland (2016) posit that the challenge with this practice is its perceived insensitivity to the plight of the dead who are disturbed during the "exhumation" processes. Secondly, most relatives of the deceased love visiting the graves of their loved ones and reusing the plot would interfere with such visits. Such practices, in some quarters, would be seen as an investment, and those buried, especially by the local governments for lack of anyone claiming them, would sometimes have their plots resold even before the prescribed time period lapses as no one would forward objections (Goodyear 2015). The final answer to these questions would be addressing graveyard used for multiple purposes besides being an exclusive place for burying the dead. Skår et al. (2018) express that such could be used as green spaces, recreation centers and jogging tracks among many others. But this is seen as a contentious stance, as posited by Harvey (2006), who explains that it would be hard to have a universal agreement on such issues-noting the push and pull between varying religious beliefs and from relatives for the deceased and with governmental policies on such practices.

3.2 Graveyards and Green Space

In most cities, graveyards and cemeteries present sites to behold. Most possess rich and deep cultural heritage, especially in regard to art and creativity denoted by the numerous, unique mausoleums (Fig. 3.2), tombstones (Fig. 3.3) and rolling and unique landscapes (Fig. 3.4) among many other things (Beard and Burger 2017; Bhatt 2016). In addition, most of these sites, especially those that are regularly manicured and kempt, represent among the greenest spaces in cities. But, unfortunately, the beauty and magnificence of these large tracts of land that occupy most of the urban centers cannot be freely enjoyed as these are not classified as regular, or acceptable, open spaces that can be freely used by the local communities. These as noted by Swensen and Skår (2018) attracts intricate and complex social, legal, environmental and political dilemmas, especially those related to morality and ethics.

From a psychological perspective, graveyards in their essence are primarily deemed as tools that help those left behind to process emotions tied to retaining emotional bonds with the deceased. So, buying from this, graveyards cannot be seen open to the broader public but should be let free for those who are bereaved. Skar and Nordh (2018) explain that the practice of visiting the dead is solemn and should be encouraged, and those visiting should be given maximum privacy and peace to mourn as they like. Therefore, opening the gates for the public would mean imposing some rules and regulations that would not augur well with those genuinely visiting gravesites. According to Bagwell et al. (2012), equating a graveyard, due to their

Fig. 3.2 Emperor Qinshihuang's Mausoleum, Xi'an, China. *Source* Manoj Kumar Kasirajan

mare size, artistic features and attractiveness and the green in them to open public spaces would be discriminatory and non-inclusive to people that do not practice burial, or to unreligious people. This thus introduces the concept that if we are to use a graveyard as an open public space, this will mean that those public spaces are reserved to only a select group of people. This is so noting that local governments would be hesitant to reserve other spaces, more so within a respectable distance to this to also serve as open public space. This would mean that those who, in one way or the other do not associate with the dead would need to either travel some distances to find other greener places they are comfortable with or remain without one altogether. On a different note, there is an emerging literature that contests the utilization of graveyards as public open spaces, as the functions associated to both cannot be made to work in a healthy manner. Quinton and Duinker (2018) showcase that, despite the myriad of benefits that can be derived from using graveyards beyond burying, risks such as groundwater degradation, availability of invasive and dangerous or poisonous species among some issues that make them unattractive in cities. Oliveira et al. (2013) explain that some of the materials used for graves and gravesites, especially those used in embalmment are poisonous and pose health risks to those who contract them.

But far from the health risks that may confront those who frequent the gravesites as open public spaces, the unhealthy relationships of using graveyards for other purposes is amplified by the emotional attachment that the bereaved have with the deceased. This is because when cemeteries are also used as public open spaces, they attract activities and ceremonies like weddings, groups' fun activities, hiking and many others that are associated with fun, joy, and contentment. But, as Huggins and Hinkson (2017) note, graveyards give the loved ones opportunities to grief and mourn privately, but in public, and it would sound absurd to have a group enjoying in one

Fig. 3.3 Père Lachaise Cemetery, Paris, France. *Source* Fabrice Nerfin

Fig. 3.4 The overcrowded the Chai Wan cemetery in Hong Kong. *Source* Sam Balye

corner while another or some individuals are grieving just within. Another issue of concern that demonstrates why graveyards should not be converted to open urban places is the issue of cost of living in some areas surrounded by some of these graves. Davis and Alexander (2008) explain that some cemeteries, surprisingly have resulted to prices of land and property adjacent to them to become exorbitant, such that, the poor, young professionals and those figuring low in the economic ladder cannot afford to live in such places (Hochstenback and Musterd 2018; Steinmetz-Wood et al. 2017). In most cases, such groups represented the highest demographic group that would benefit from the availability of open spaces in their areas. Unfortunately, with the cost of living beyond their reach, they are relegated to further neighborhoods that may be at distances further from the graveyard purported to be the open green space for the city. Another issue as noted by Harvey (2006) is that a sizeable number of graveyards are associated or owned by religious groups; hence, perceived as sacred places; hence, using them as "park" can often seem blasphemous to some, and such would spark conflicts fueled by religious arguments. To further explore how graveyards operate and the challenges and perhaps opportunities it brings to cities, three case studies are explored in the next section.

3.3 Case Studies

3.3.1 Coleman Cemetery

Coleman Cemetery in Alexandria is another worth case that helps showcase the impact of having graveyards in urban areas. According to Lee Yang, this cemetery occupies three acres of land and approximately 1,200 burials (Tang 2019). Nevertheless, its size and relatively small number of burials have shown lacked in planning shown by larger cemeteries occupying thousands of acres and hosting millions of burials. Like them, it has also been somehow neglected. Reasons cited, like in other cases include having a large population of unnamed (or unclaimed) buried; often being poor, homeless and who lived in destitute conditions, and with no one stepping forward to pay for a dignified burial for them. In this case, most of those are buried by W. H. Bacon Funeral Service; a private agency contracted by the local government to dispose all unclaimed bodies from within the cities. According to McCoy (2019), the contract between the local government and this agency entailed, among other things, the need for "a permanent marker" at the gravesite or mausoleum for each buried body, bearing proper identification of the person. Nevertheless, this is argued as not being the case and it is claimed that some were first cremated, and their ashes buried in a mass grave with no names or tombstone for that matter (Davis 2015).

From a moral perspective, this may indirectly mean that the person was unloved both in the living and in death, thus, no one cared whether the place they were put to rest is kempt or not. This practice can be both moral and ethical arguments. Nevertheless, such trends of burying the unclaimed in a mass grave are seen as being more practices to both the contracted funeral home and to the local government (Tang 2019). In the case of the local government, it means that the demand for more land for graveyard would be slowed down; hence, save them the trouble of having to search for a new piece amid the scarcity. Similarly, it saves on not having to continuously maintain the cemetery as there no many people related to the dead would visit and feel aggrieved of the neglect (Davis 2015). The funeral home benefit by saving on costs of buying caskets, tombstones, emblements and other associated miscellaneous. On a negative note, such practices have the potential to discourage maintenance of the cemetery, and this would spiral into rapid urban decay of the area.

The sorry state of affairs at Coleman are amplified by the state of Snowden & Bethlehem Cemetery that is located just next which looks much better, and as Tang (2019) noted, the prices of houses there were reasonably higher; $1.2 million for a house as opposed to the area average price of slightly above $500,000 as reported by (Zillow 2019). The former is said to be maintained by Bethlehem Baptist Church working hard to ensure that each gravesite is clearly marked, and the environment therein cared for. For Coleman Cemetery, Davis (2015), argued that there was only one caretaker for this grave; an old man named Roy Reave, who was also perturbed by the state of affairs in the cite, especially in regard to unnamed and unmarked graves. The maintenance of Snowden and Bethlehem Cemetery, hence its distinct difference with Coleman shows how religion plays a vital role in issues pertaining burial, and how the same can be relied upon in making such sites more coherent with urban demands and dictates.

3.3.2 Rock Creek Cemetery

Rock Creek Cemetery occupying approximately 86 acres piece of land is another prominent and historical site that showcases how cemeteries located in urban areas have direct impact of its fabric and its reputation. Established in 1719 by the British colonials under the St. Paul's Episcopal Church, Rock Creek Parish to serve as the church graveyard, the cemetery was later open to also serve the public as from 1791, and in 1840, under an Act of Congress, the cemetery was formally established (Paul's Rock 2019). It is natural rolling landscape and beautiful countryside garden style gave it prominence to serve also as a public park. To complement its natural orders, the cemetery is rich with numerous, notable and historical artworks ranging from mausoleums, sculptures, and tombstones that prompted the cemetery and the churchyard to be enlisted on the National Register of Historic Places. The prominence of this cemetery attracted attention, and for that reason, many prominent national figures from diverse backgrounds and occupation are buried there. Judkis et al. (2018) posit that to cap it all, this cemetery is home to statue of "Grief", one of the famous sculptures in Washington DC, marking the gravesite of Henry and Maria Adams, which was commissioned by Henry who contracted a renowned sculptor; Augustus Saint-Gaudens to honor his wife who committed suicide by drinking painting chemical. The sculpture has been a source of attraction to many.

Despite its rich history, it is said to be slowly being left to despair and a number of reasons are credited to this trend. First, it is noted with concern that its water bills have been skyrocketing at an alarming rate. According to Fleischer et al. (2019), it is said that this rose from an annum average of $3,500 to a high of $200,000 per annum. The rise is said to have been prompted by DC Water charge that is based on the estimated amount of surface runoff that each particular property produces; thus, polluting the Anacostia River. The cemetery charges are based on the total surface different mausoleums occupy in the property. The local government's intention to clean the river, though hurting to the property owners, including the cemetery are justified, since, a cleaner river make the area around it attractive and contribute to sustainable efforts.

The controversy between the Rock Creek Cemetery and the municipal council can be resolved by adopting the approach by the management of Mount Olivet Cemetery where they allowed surface runoff from other areas to be diverted into the cemetery (Hickman 2018). This allowed for reduction of pollution runoff from reaching the Chesapeake Bay, while at the same time, helping in keeping the environment greener, and even better, allow the Mount Olivet Cemetery management safe over 4% of their annual water bills. By adopting a similar approach, Rock Creek Cemetery would also manage to reduce some of the water bills, and improve the quality of environment, making it even more attractive as mixed-use piece with cemetery and recreation park services. By adopting this approach, the cemetery could also help the local community safe on their water bills tool by allowing them, through the municipality to have their surface runoff also diverted in its massive piece of land that can accommodate evil drop flowing from the neighborhood. This would allow the river cleaning process to be successful; hence, contribute to ensuring the resident have enough water for their use. In this case, the economic sustenance of the cemetery as a singular activity will not be justified, and the property owners will ultimately be forced into the development of the site to other uses.

3.3.3 Holy Rood Cemetery

The last case study pertains to the Holy Rood Cemetery that was established back in 1832 as Holy Trinity's parish cemetery in a six-and-a-half-acre piece of land, which presents an opportunity to understand the impact of cemeteries in growing urban areas. From its history, it is recorded that the cemetery was the third after the Holy Trinity Church graveyard and the College Ground that were established earlier. This cemetery was maintained by parish through the sale of burial spaces, but as of 1915 when the last such plot was sold, financial constraints were felt (Kelley 1795). Such caused visible cases of neglect that angered relatives and other stakeholders who raised funds to care for it, but only succeeded for a while. For this reason, the parish partnered with Georgetown University, which was also partly owned by the church to care for the cemetery. But this still did not solve the neglect cases, and it is recorded that, in summer, during mowing, most headstones were scarred (Fletcher 2002). The stones are also disfigured, tumbled or displaced by fallen trees or by tracks hauling the trees away. The neglect is also amplified by the fact that there are no caretakers or security patrols to prevent intruders and vandalizes. According to Fletcher (Fletcher 2002), the reason behind this neglect, beyond money, is the intention by management of the University to have the land reused after 100 years, especially in commercial real estate developments and by 1972, even before the lapse of the 100 years, the university was ready to kick start the process of reclaiming the cemetery.

According to Gillis (2019), the disinterment of the bodies and reburial, in Mount Olive Cemetery which had been identified as the relocation site was to be done by the university, since the church did not want to cause distress to the community, which was, from the start controversial. By the time the university was ready for the exercise, the church, and the local community objected the idea in its entirety. The university was motivated by the real estate potential that the piece of land had, and by getting the go-ahead, it would have helped solve the social issues, especially housing that is already evident in this neighborhood. But, noting that there was a score of slaves that were buried in the cemetery, their relatives did not want anything to do with the relocation (Fletcher 2002).

The contestation demonstrated in this case study showcased how intimate the issue of religion and death are intertwined, and how people feel connected to the dead, even after centuries. The same demonstrates how cemeteries have the potential to hold back much-required development opportunities in cities. When unkempt, like in the case of Holy Rood Cemetery, it highlights how such has the potential to spur insecurity, reduce the attractiveness of the place, and ultimately, this affects the price of available properties.

3.4 Graveyards and Inclusivity

Noting the emerging issues from graveyards as depicted in the case studies above, one cannot help to wonder about a number of issues. The first issue that one struggles with is that of whether graveyards and cemeteries can stand as truly inclusive spaces for surrounding neighborhoods. Second, whether such burial places need to remain as exclusive areas for the bereaved to process emotions, grief and mourn as they try to retain their connection with the deceased. On the first issue, the words of Seng (2012) are strong, especially basing his argument on religious diversity and the perception of such religious outfits concerning how the dead should be disposed of. According to him, each group is entitled to their beliefs, but there ought to be some form of equity. That is, those who are pro cremation should not be seen to suffer from actions and inactions of those who practice the religious act of burying the dead. This is more so in respect to the diminishing space for human settlements as cemeteries and graveyards, some inactive for years, are made to occupy massive tracts of land and some in very potential places for mixed-use development (Ryan 2017); as is shown in Fig. 3.5 showcasing the Trinity churchyard cemetery which was quickly found to be at the centre of a prime spot in the city of New York; where the American Stock

Fig. 3.5 Trinity Churchyard cemetery in front of the American stock exchange building in New York. *Source* Jahsie Ault

Exchange building sits. The question arises of how far this cemetery can be kept as it fights under the sprawling forces of the city and that of urban economics.

A similar case can be found in that of the Holy Rood Cemetery, where the Georgetown University was interested to even go further and approve the disinterment bodies and transfer them to another cemetery, and on their place, build commercial and residential developments that would serve the larger population. But, on ethical grounds, their decision was pulled down, especially with the church distancing itself from such practices. If the plan was actualized, the real estate ventures would have brought some level of equity on the use of that public land. But, since that failed, like is the case of many places where there is contestation between religious beliefs, moral and ethical principles and the local authorities proposals to introduce the concept of mix-use of cemeteries, people are forced to give way for religious arguments tied to the singular use of cemeteries (Bennett and Davies 2015; Löki et al. 2019). Mostly, such relocations are triggered by gentrification which occurs as the few remaining spaces and older parts of the cities are renovated or undergo extensive real estate investments that the poor cannot put up with (Larsen et al. 2010). As this happens, the dead, who are buried right within the centers of the city are seen to be occupying prime real estate over a majority of the living; who are relegated to sprawled areas and made to experience numerous challenges impacting negatively on their livelihoods.

On the second scenario of whether burial places should remain exclusively reserved for the dead and those who genuinely visit the place for emotional and psychological healing; despite the moral and ethical perception supporting that it should not be the case. Numerous studies have showcased that there are numerous benefits that the people, the city and the economy at large can derive from using, or reusing, cemeteries or graveyards for other purposes (Neckel et al. 2017). According to Quinton and Duinker (2018), graveyards serves as a buffer zone for erosions, surface runoffs and other extreme weather conditions that could have far-reaching impact if they were to directly impact urban areas. This is demonstrated in the case of Mount Olivet Cemetery that is being used to absorb the surface runoff from areas neighboring it; hence, helping in reducing the amount of pollutions and other negatives that are caused by runoffs (Hickman 2018; Rogers 2019). On the same front, Clayden et al. (2018) explain that cemeteries have numerous ecological benefits, like serving as a home for rich wildlife and ecosystem for numerous birds, animals, insects and wide range of plant species that are disturbed, destroyed or displaced by human activities, especially urban sprawl. From these examples, denoting graveyards as exclusive places for the dead and their loved ones would mean that such benefits are compromised, as we would not be able to make those areas more ecological rich and diverse, like in the case of Holy Rood Cemetery that is said to have been neglected due lack of finances to maintain and fund some activities that would make it align with urban environment.

This issue of finance is very sensitive, especially if the funds are sourced from the public coffers. If those burial places are seen to be servicing only by an exclusive segment of the society, then how can the local government justify its use of public money on a property that does not satisfy the public at large? Such trends would give those who feel excluded from using the public open space the impetus to question the local government and put them to task on their accountability with their use of public resources. Such would also reveal loopholes on governance practices and would raise more challenges than solutions; hence, the issue of exclusivity of those areas needs to be carefully examined.

But, even as we question the viability of graveyards as exclusive places for the dead and the believed, we need to acknowledge the role that such places play in emotionally assisting people to come to terms with the reality of death and heal from the wounds inflicted by human losses (Gorer 1965; Klass and Walter 2001). We also need to appreciate and acknowledge the religious convictions of a majority who choose to bury their dead and advocate for the graveyards to remain as sacred places for the dead. But, while still on this, we also need to question the moral and ethical imperatives of what such actions of accepting graveyards as exclusive places and ask whether there could, or would, potentially be alternative solutions that could help address this conundrum.

3.5 Conclusion

This chapter introduces an emerging issue arising from the religious practice of burials regarding the processing of the dead and explores how its practice has been ingrained in our societal mannerisms and can aid in the emotional processing human loss, helping in attaining closure and in keeping a connection with the dead. However, as the psychological and emotional literature supports this practice, we see the emergence of urban forces; mainly due to increasing urbanization coupled with that of a demographic boom that dictates that graveyards are being seen as occupying vast spaces of land for the service of only a handful of people at any given time. This is aggravated by the fact that due to sprawling cities, graveyards are now considered to be at the center of cities-often victim to gentrification, and where people have to relocate to other areas, whereas the dead are enjoying prime spots in urban centers. This then asks us to pose the question as to the sustainability of this practice and its ethical and moral imperatives.

References

Bagwell S, Evand G, Willing A, Worpole K (2012) Public space management. Report to the intercultural cities research programme; Cities Institute, London

Beard VR, Burger WC (2017) Change and innovation in the funeral industry: a typology of motivations. OMEGA J Death Dying 75:47–68

Bennett G, Davies PJ (2015) Urban cemetery planning and the conflicting role of local and regional interests. Land Use Policy 42:450–459

Bhatt K (2016) Cemeteries as healing landscapes. Unitec Institute of Technology, New Zealand

Chapple K (2016) Income inequality and urban displacement. New Labor Forum 26:84–93

Clayden A, Green T, Hockey J, Powelld M (2018) Cutting the lawn—natural burial and its contribution to the delivery of ecosystem services in urban cemeteries. Urban Forest Urban Green 33:99–106

Davis S (2015) This is what happens to unclaimed bodies in Washington, DC. https://www.vice.com/en_au/article/9bze43/this-is-what-happens-to-unclaimed-bodies-in-washington-dc-406. Accessed 15 June 2019

Davis H, Alexander L (2008) Karl would not be amused. The Times

Fleischer J, Yarborough R, Jones S (2019) Historic D.C. Cemeteries in jeopardy of closing due to growing water bills. https://www.nbcwashington.com/investigations/Historic-DC-Cemeteries-in-Jeopardy-of-Closing-Due-to-Growing-Water-Bills-452918463.html. Accessed 15 June 2019

Fletcher C (2002) Burial grounds of holy trinity church, Georgetown, DC. Newsletter of the Catholic Historical Society of Washington, vol 10

Gade DW (2015) Cemeteries as a template of religion, non-religion and culture, pp 623–647

Geisler NL, Potter DE (1998) From ashes to ashes: is burial the only christian option? Feature Article ed.; Christian Research Institute: Charlotte, NC

Gillis J (2019) Holy rood: a cemetery with a tell-tale heart. https://www.thehoya.com/holy-rood-a-cemetery-with-a-tell-tale-heart/. Accessed 15 June 2019

Goodyear S (2015) If you die poor or alone in New York city, you'll be buried by prisoners in a mass grave. https://www.thenation.com/article/there-progressive-way-bury-lonely-poor-huddled-masses/. Accessed 5 July 2019

Gorer G (1965) Death, grief and mourning in contemporary britain. Cresset Press, London

Harvey T (2006) Sacred spaces, common places: the cemetery in the contemporary American city. Geogr Rev 96:295–312

Hayward SF (2018) Colonial expressions of identity in funerals, cemeteries, and funerary monuments of nineteenth century perth, western australia. Genealogy 2:1–26

Hickman M (2018) Historic dc cemetery doubles as pollution-absorbing sponge. https://www.mnn.com/green-tech/research-innovations/blogs/historic-dc-cemetery-doubles-pollution-absorbing-sponge. Accessed 15 June 2019

Hochstenback C, Musterd S (2018) Gentrification and the suburbanization of poverty: changing urban geographies through boom and bust periods. Urban Geogr 39:26–53

Huggins CL, Hinkson GM (2017) Contemporary burial practices in three Caribbean islands among Christians of African descent. OMEGA J Death Dying 1–14

Johnson TM, Zurlo GA, Hickman AW, Crossing PF (2018) Christianity 2018: more African Christians and counting martyrs. Int Bulletin Mission Res 42:20–28

Judkis M, Hahn F (2019) Creeped out by cemeteries? These DC Sites—with history, art and even yoga—might change your mind. https://www.washingtonpost.com/goingoutguide/creeped-out-by-cemeteries-these-dc-sites–with-history-art-and-even-yoga–might-change-your-mind/2018/10/24/9344a0ca-d2d7-11e8-83d6-291fcead2ab1_story.html?utm_term=.739e5ac3bea7. Accessed 15 June 2019

Kelley LJ (1945) The history of holy trinity parish, Washington, DC 1795–1945. John D. Lucas Printing Company: Baltimore

Klass D, Walter T (2001) Processes of grieving: how bonds are continued. In: Stroebe MS, Hansson RO, Stroebe W, Schut H (eds) Handbook of bereavement research. American Psychological Association, Washington, D. C, pp 431–448

Larsen J, Coleman WJ (2010) Cemetery proximity and single-family house price, vol 78, pp 33–49

Löki V, Deák B, Lukács AB, Molnar VA (2019) Biodiversity potential of burial places—a review on the flora and fauna of cemeteries and churchyards. Global Ecol Conservat 18:e00614

Mathijssen B (2017) The ambiguity of human ashes: exploring encounters with cremated remains in the netherlands. Death Stud 41:34–41

McCoy T (2019) Washington's unclaimed remains in unmarked graves next to trash cans. https://www.washingtonpost.com/local/social-issues/washingtons-unclaimed-remains-in-unmarked-graves-next-to-trash-cans/2015/07/19/a8c03ce2–24df-11e5-aae2-6c4f59b050aa_story.html?noredirect=on&utm_term=.aa55a6a40d41

Neckel A, Costa C, Mario DN, Sabadin CES, Bodah ET (2017) Environmental damage and public health threat caused by cemeteries: a proposal of ideal cemeteries for the growing urban sprawl. Urbe. Revista Brasileira de Gestao Urbana 9:216–230

Niţă MR, Iojă IC, Rozylowicz L, Onose DA, Tudo AC (2014) Land use consequences of the evolution of cemeteries in the bucharest metropolitan area. J Environ Plann Manage 57:1066–1082

Oliveira B, Quinteiro P, Caetano C, Nadia H, Arroja L, da Silva EF, Matias MS (2013) Burial grounds' impact on groundwater and public health: an overview. Water Environ J Promot Sustain Solut 27:99–106

Quinton J, Duinker P (2018) Beyond burial: researching and managing cemeteries as urban green spaces, with examples from Canada

Rogers J (2019) Planning for a rainy day: Washington, DC, combines regulations with a new credit market to keep polluted stormwater out of its rivers. Nat Conserv. Accessed 1 Mar 2019

Rotar M, Todor I (2013) An investigation about the attitudes towards death and cremation in contemporary Romania. Soc Behav Sci 82:192–197

Rugg J, Holland S (2016) Respecting corpses: the ethics of grave re-use. Mortality 22:1–14

Ryan K, Steinmetz C (2017) Housing the dead: What happens when a city run out of space? https://www.abc.net.au/news/2017-01-06/what-happens-when-cities-run-out-of-space/8165284. Accessed 5 July 2017

Seng LK (2012) History of the dead, heritage of the living: Bukit brown cemetery in Singapore. https://kyotoreview.org/issue-12/history-of-the-dead-heritage-of-the-living-bukit-brown-cemetery-in-singapore/. Accessed 4 July 2019

Shepkaru S (1999) From after death to afterlife: Martyrdom and its recompense. Assoc Jewish Stud 24:1–44

Skar M, Nordh H (2018) Swensen G. More than just parks, Green urban cemeteries, pp 1–21

Skår M, Nordh H, Swensen G (2018) Green urban cemeteries: more than just parks. J Urban Int Res Placemaking Urban Sustain 11:362–382

St. Paul's Rock Creek Episcopal Parish (2019) Our history. https://www.stpaulsrockcreek.org/our-history/. Accessed 15 June 2019

Steinmetz-Wood M, Wasfi R, Parker G, Bornstein L, Caron J, Kestens Y (2017) Is gentrifcation all bad? Positive association between gentrifcation and individual's perceived neighborhood collective efcacy in Montreal, Canada. Int J Health Geograph 16:24–33

Swensen G (2018) Between romantic historic landscapes, rational management models and obliterations—urban cemeteries as green memory sites. Urban Forest Urban Green 33:58–65

Swensen G, Skår M (2019) Urban cemeteries' potential as sites for cultural encounters. Mortality 24:333–356

Tang J (2019) Cemeteries use a lot of space and are terrible for the environment. Is there a better way? https://ggwash.org/view/70300/burial-culture-and-the-issues-with-using-so-much-space-for-cemeteries. Accessed 15 June 2019

Tudor CA, Iojă IC, Hersperger A, Pătru-Stupariu I (2013) Is the residential land use imcompatible with cemeteries location? Assessing the attitudes of urban residents. Carpathian J Earth Environ Sci 8:153–162

Zillow (2019) Fairfax home prices & values. https://www.zillow.com/fairfax-va/home-values/. Accessed 15 June 2019

Chapter 4
Theology, Sustainability and Big Data

Abstract The subject of faith has been observed to emanate in view of uncertain dimensions that impact negatively on psychological and liveability levels of individuals or of collective societies. However, as the amount of data increases from contemporary urban planning concepts such as Smart Cities, there are increasing tools for forecasting and predictions based on machine learning and artificial intelligence that can provide precise and accurate information; thus, reducing unknowns and uncertainties. While those are seen to increase the efficiency and performance of urban fabrics, there is an equal understanding that those can be made to impact on faith, and consequently on the governance of cities; which are shown to be intricately linked with religious principles and ethos. The subject of faith, data and governance thus emerges and is shown to be of larger significance as the role of technology is made more pronounced in urban areas.

Keywords Theology · Sustainability · Big data · Technology · Climate change · Resilience

4.1 Introduction

Human beings are in constant pursuit of higher powers or forces, to whom they seem to trust and rely. This faith is based on a myriad of factors, which are partly influenced by individuals or group experience on diverse aspects of life. Among these factors is the fear of unknowns which have been discoursed and analyzed widely and has to elucidate diverse viewpoints, especially on how people tend to cope (Nath 2015). It has been argued that one of the commonest way that people use to circumvent the notion of fear is that of seeking companionship with supernatural forces (god(s)) whom they perceive to have powers over the causes of these fears. Oman (2018) argues that this tendency points to the psychological reaction to fear of diverse things; which Carlisle (2013) characterizes as being fear of "the whole thing-mysterious, defeat, diseases, death and afterlife among many other things. This argument was explained by Russell (1927), who candidly posited that "religion is based primarily and mainly upon fear and this drives individuals to seek and trust a perceived higher power that is omnipresent even in the times of unknown and uncertainties.

Z. Allam, *Theology and Urban Sustainability*, SpringerBriefs in Geography,
https://doi.org/10.1007/978-3-030-29673-5_4

In the wake of climate change and its immense impacts, the fear of unknown is seen as though it has been accelerated, especially among the urban dwellers. This is so especially noting that there are now increased cases of flooding (Levitt and Kommenda 2018; European Commission 2018; EPA 2018; Allam 2012; Allam and Jones 2018), heat waves, reduced resources like water and other unpredictable happenings like earthquakes among others (Crist 2007; Goldthau 2014). It has also been noted that with climate change, there are increased cases of draught, change in weather patterns and rise of new traits of diseases and sickness (Ogden 2018). Short et al. (2017) decry that climate change is causing incidences where the emergence and re-emergence of parasitic diseases that are being experienced globally. Henderson et al. (2018) claim that this resulting disruption on our lifestyle is also, in part, responsible for unpredictable business environments with economies baring the blunt of reduced job opportunities and reduced financial resources to fund government and private projects; thus, increasing burdens of debt and other related economic challenges. The impacts of climate change are thus apparent on urban as well as national economies, and even on superpowers such as U.S. especially due to their impacts on national security, which is said to be threatened by increased natural disasters, immigrant crisis and the global increase of conflicts over basic resources (Department of Defense 2015).

When all these fears are compounded, it has been found that most people turn to religion and diverse faiths, which are deemed to give solace and hope. While on this, it has been linked that in times of fear, most places of worship receive unprecedented numbers of worshippers and ultimately increased offerings, both financial and in kind as people attempt to appease the gods with a hope of getting closer to them for redemption especially when the unknown strikes (Leigh and Hambler 2014; Boyer 2003; Markham 2012). This notion even explains why the number of religious adherents is substantially higher globally than in some years back. For instance, Sherwood (2018) estimate the number of religious adherents to be approximately 84% of the global population in 2018, which has been increasing substantially as shown in Table 4.1.

These uncertainties, however, are not only addressed in faith-based arenas, but also in the scientific front, which has gained traction, especially due to technological advancement. In the recent past, there has been a noticeable trend denoting how these uncertainties are being pursued and rationalized by scientific breakthroughs, and thus, leading to a more secular society. In this regard, the advancement of science is leading to more technologically rich urban fabrics that are producing a vast array of data. In the forefront, through advanced technologies, scientists are now able to propose and implement a plethora of strategies aimed at mitigating the impacts of climate change as is noted in a report by the United Nations Framework Convention on Climate Change (UNFCCC 2006). In urban areas, the employment of such technologies like the Internet of Things (IoT), Artificial Intelligence (AI), Machine Learning, Big Data, and Cloud Computing among many others are all geared toward ensuring sustainability, and more so against threats posed by climate change. The data generated via these technologies and the installed smart components (as promoted by Smart City concepts) allow scientists to accurately predict the probability

Table 4.1 World religions by adherents, 1910–2010

Religion	Adherents 1910	%1910	Adherents 2010	%2010	1910–2010% p.a.	2000–10% p.a
Christians	611,810,000	34.8	2,260,440,000	32 8	1 32	1.31
Muslims	221,749,000	12.6	1,553,773,000	22.5	1.97	1.86
Hindus	223,383,000	12.7	948,575,000	13 8	1.46	1.41
Agnostics	3,369,000	0.2	676,944,000	9.8	5.45	0.32
Chinese folk religionists	390,504,000	22.2	436,258,000	6.3	0.11	0.16
Buddhists	138,064,000	7.9	494, KB 1,000	7.2	1.28	0.99
Ethno religionists	135,074,000	7.7	242,516,000	3.5	0 59	1.06
Atheists	243,000	1.0	136,652,000	2.0	6.54	0.05
New religionists	6,865,000	0.4	63,004,000	0.9	2.24	0.29
Sikhs	3,232,000	1.2	23,927,000	0.3	2.02	1.54
Jews	13,193,000	0.8	14,761,000	0.2	0.11	0.72
Spiritist	324,000	0.0	13,700,000	0.2	3.82	0.94
Daoists	437,000	0.0	8,429,000	0.1	3.00	1 73
Baha'is	225,000	0,0	7,306,000	0.1	3.54	1.72
Confucianists	760,000	0.0	6,449,000	0.1	2.16	0.36
Jains	1,446,000	0.1	5,316,000	0.1	1.31	1.53
Shintoists	7,613,000	0.4	2,761,000	0.0	−1.01	1.09
Zoroastrians	119,000	0.0	197,000	0.0	0.51	0.74
Total population	1,758,412,000	100.0	6,895,889,000	100.0	1.38	1.20

Source Johnson and Grim (2013)

of certain climatic events taking place; hence, proper planning and arrangements are made beforehand and also, in the same length, have helped in the demystification of some uncertainties that have buffed people for long. Scher and Messori (2018) explain how technologies such as deep learning, riding on convolutional neural networks are allowing scientists to use past weather patterns to predict what is to be expected many years in the future.

Besides weather predictions, scientists have also advanced in areas like production of renewable energies, production of smart automobiles and other devices that do are not necessarily powered by fossil fuel products. Also, they have managed to come up with technologies that are geared towards the generation of food in urban areas through green technologies. As noted above, all these advances are helping weather down a wide range of fears that were turning people towards religion, and by so doing, people are now questioning some of the fundamentals of religious beliefs and dogmas. These dogmas, especially in the urban areas are now been taken with "a pinch of salt" and this has even led to emergence of a sizeable group of people who believe in existence of God, but do not subscribe to any known religious outfit as was observed by Overstreet (2010).

With the role of religion in giving answers and solution to unknowns thawing down, cities around the world are now seen to embrace technologies that allow people to take more control of how different aspects of the city are done. It has been observed that, with technologies, there are new perspectives on how issues like governance, relationships, consumptions, care for environment and infrastructures are rendered. For instance, having noted that the many fears that people have are surmountable, cities are now seen to embrace construction of architectural structures that defy heights. There are increased, conscious efforts directed toward conservation, inclusivity, security, and sustainability among many others. Governments, people and other stakeholders are more concerned about the promotion of practices that would reduce population increase in cities, while at the same time, improving on the health status of those living in them, and in doing so the reevaluation of some principles, like religion and relationship with urbanism, that were previously commonly accepted need to be pursued.

4.2 Climate Change, Sustainability, and Religion

It is interesting to see how religion is fuelled by fear, and how surprisingly, most of these fears emanate from the religious teachings and beliefs. In particular, since time immemorial, different religious groups hold that the world will end in a catastrophic way, and the afterlife is to be enjoyed only by those whose actions and (where appropriate) inactions are right in the sight of the gods. It is believed that some sort of cataclysmic event will forerun the end of time, and though each religion describes it differently, these are meant to evoke fear among adherent of the particular religious group; hence, conform to the teachings and prescripts of the religion (Silton et al. 2011). Beyond the end of the world, diverse religious groups teach that there will be everlasting happiness or punishment for those who proceed to heaven or hell respectively. Christians, for instance, has a profound explanation of events that will precede the end time and those that will come after in the afterlife (Kyle 2912). For this, the last book of the Christian Bible; the Revelation, is almost dedicated to the end of time and, its message, though other books within the Bible emphasis on the same thematic. In summary, the Christian belief that Christ will return and take home His church "the believer" and those left behind will have to contend with pain, strive, violence, and all sought of evil things. After 1000 days, Christ will return and judge the earth and those found unworthy will be cast in a lake of fire (Hell) where they will spend their eternity. For the believers, they will proceed to heaven where they will spend their eternity with God (BBC 2011).

Hindus, on the other hand, believe that the end of time will be brought about by *Pralay (destruction)* instigated by a raging flood that will wipe away everything (Ashe 2001). Those in Judaism are told that the end of time will come when all those exiled from Israel return and the dead are resurrected, but of these, those found guilty of any sin will have to spend eternity in hell, which is everlasting fire. Zoroastrianism argues that an all-consuming fire will engulf the earth with the sinners burning continuously for three days, after which they will be forgiven (Boyce 2002). Like the Christians, the Muslim believes that the end of the world will be preceded by folding up of the sun, falling of stars, oceans losing their boundaries and immense noises and clamor and other horrifying events that will cause a lot of sweating and unfathomable fear (Al-Ghazali 2016).

Interestingly, as explained by Roser-Renouf et al. (2016), the recent advents of climate change are being equated to those foreseen cataclysmic events that will forerun the “end of days”. According to them, most people perceive the issues of climate change as the works of God as he tries to forewarn the end of days and thus, people should not be concerned on taking actions against these events as it is not in their power to change things. This proposition is affirmed by Barker and Bearce (2013) who found out that most end-time Christians held that the coming back of Jesus was nigh, hence, found it unnecessary to worry about issues like climate change. In fact, they base their argument on the biblical teachings that explain that before the end, such things like famine, violence, and environmental degradation will inevitably be present, as explained by Williams (2008). There is this thus conviction that the challenge of climate change (equating to the end of days) can only be overcome through fervent prayers and addressing it can only be pursued by the gods, not humans (Schuman et al. 2018). Hulme (2009) posits that Christians are commissioned by God to take care and have dominion over his creation. Therefore, when such challenges like climate change strike, it is viewed as a punishment from God and it is only Him who can reverse it when beseeched in prayers and supplications.

Despite their religious convictions, non-end time believers, does not attribute actions against climate change to godly events since scientists have already addressed this by providing practical explanations on the causes of climate change (Jooste et al. 2018; Burkart 2017). This is accentuated by visuals which speaks faster to our current connected world and images as shown in Figs. 4.1 and 4.2 readily speaks to a large audience. Similarly, they have also extensively provided practical and collective actions that can be adopted in different scales to arrest the situation (Seidler and Stevenson 2017; Stecula and Merkley 2019). Indeed, the intensity of the scientists’ works are evident through the different conventions that have organized at different periods, all geared toward encouraging countries, and government to adopt practices that have the potential to reduce these impacts. The latest of such conference, the COP24, was held on December 2018 in Katowise, Poland, and the emphasis, like in other previous meetings was on addressing issues that linked with climate change (UNFCCC 2018), however, the link to religion remains absent due to the sensibility of the subject. Something noteworthy on this front is that the climate change agendas have been observed to be championed mostly by the youthful generation, with most of the older people, even those in positions of authority taking the back seat (Winston 2019; Karol 2018).

The youth is seen to be leading this endeavor, since, most of them advocate for the pursuit of sustainability through actions that are imprinted in concrete policies that

Fig. 4.1 Urban area flooded. Photo by Chris Gallager

Fig. 4.2 Drought and desolation in Kota Garut, Indonesia. Photo by Dikaseva

have the potential to guide these actions to the furthest future (Farber and Reichert 2018). Though this is the case, the tragedy that the environment is facing is that most of the influential leadership positions are held by the older generation, who are more susceptible to religious beliefs (Skirbekk et al. 2018), and hence; encourage inaction on those issues. A report by Reinhart (2018) shows that in America for instance, 56% of people aged 55 years and above were in support of climate change action as compared to over 70% of those aged below 35 years. A similar report by Jones et al. (2014) revealed that over 46% of senior citizens (who represent a majority of those who hold positions of leadership) regard high their religiosity compared to only 29% of young people, and as noted, a majority of religious people attribute issues of climate change to God, while those less religious or non-religious attribute it to human activities and behaviors.

Despite the ambiguities in addressing the challenges associated with climate change, it is doubtless that leadership must emanate from governance structures that are secular in beliefs and oriented towards science and practicality. This though does not take away from religion its role in the society, but, scientific experience has shown that some of these challenges, unlike religious explanations, are as a result of human actions. Therefore, on this, relying on religious teachings and emphasis on prayers does not serve the intended purpose of regeneration, while adopting scientific parameters, guided by firm policies, on the other hand, would allow for a reversal of the situation and help remediate on the harm caused by climate change.

4.3 Big Data, Governance, and Religion

The advent of Big Data, coupled with other technologies, has brought solace, especially in respect to the numerous uncertainties that have confronted the human race, as advanced by Faghmous and Kumar (2014). It is evident now that a majority of unknowns can effectively be reduced through the use of data, and this is more accentuated in urban areas where technologies such as smart city concepts are being implemented and promoted (Allam and Newman 2018; Allam 2018). To put this into perspective, it is worth noting that Smart City technologies rely heavily on diverse sets of data from numerous smart sensors, devices, and systems among others smart component(Allam 2019; Allam and Dhunny 2019). Among these data, are those related to health and environment generated from sectors like the energy generation, telecommunication and the construction industries that had for long sparked fear due to the amount of emissions they were reported to be producing. Bose (2010) explains that some of these sectors have been on the forefront in generating substantial amount of dangerous pollutants that in turn have contributed highly in perpetuating the effects of climate change. Nevertheless, with smart technologies, like machine learning (Mahdavinejad et al. 2018), AI (Wright and Schultz 2018; Tzafestas 2018) and IoT (Čolaković and Hadžialić 2018; Patel and Patel 2016) among others, scientists have managed to instigate measures, through data that are geared toward promoting practices that lead to minimum generation of these pollutants while optimizing resource use and outputs (Manogaran and Lopez 2018; Singh and El-Kassar 2019).

Besides climate change, these technologies have given scientists the impetus to pursue answers to big questions that have for long been puzzling societies. For instance, through advanced technologies, people like Prof. Hawking are known to have attempted to ponder on the existence of god and the link between space and time (Hawking 2001). Similarly, technology has allowed organizations such as NASA to explore other planets and in this course, attempt to establish whether there exist extraterrestrial beings or life beyond the atmosphere of our planet (NASA 2018). On other fronts, technology is now allowing scientists to devise systems like Artificial Neuron Networks (ANNs) (Zorins and Grabusts 2015; Bryck and Fisher 2012), which, in the fullness of time, are projected to have the capacity to mimic neuroplasticity (capabilities of the human brains); hence, allowing for urban systems to function with minimum human interaction (Sasmita et al. 2018; Huang 2017; Zorins and Grabusts 2015; Kotenko et al. 2015). In the recent past, for instance, through such technologies, there have been a noticeable increased usage of non-human object like robots, drones and other specialized non-assisted technology that requires minimum human interventions to perform complex tasks at phenomenal speed and accuracy, even to the point of creating envy by humans (Mahdavinejad et al. 2018; Čolaković and Hadžialić 2018; Bačić et al. 2018; Deshpande et al. 2016; Kumar et al. 2012; Sood et al. 2018). The success of such projects will further demystify some of the religious dogmas that have limited humankind's ability in taking advantage of the various tools at our disposal to help mold a better society.

The pursuits, as initiated, by Prof. Hawkings and NASA among others are just a tip of the iceberg when it comes to disapproving some religious, long-term beliefs that have captivated the mind of numerous religious adherents and kept them in a state of continuous fear. The increased use of data is already perceived as an interesting phenomenon on this front as it allows scientists to decode some of the religious beliefs like the causes of climate change that cannot hold henceforth (Schuman et al. 2018). Indeed, it has been found that science is even gaining substantial grounds even in religious discourse. This is accentuated by increasing amounts of data that is made to allow more scientists to come forward with new theories that explain the nature of the world and helps in demystifying complex processes and entities, which may be seen as contradictory to certain religious beliefs; hence, creating disparities between science and religious discourses. One such discourse is on the form and shape of the earth, and which between them rotates around the other. Some critics accused a number of religions of advocating that the earth was flat with the sun rotating horizontally around (Garwood 2007; Schadewald 1979). Nevertheless, scientific research established that the sun is spherical and, together with other cosmic bodies, rotates around the sun. Using modern technologies like aerospace photography and space exploration, numerous scientists have proven this fact. Today, despite a few groups who are not necessarily religious still holding ground on the flat earth theory, the scientific explanation is almost conventionally accepted, with religious groups having had to change their text to this reality (Ayala 2008).

Borrowing from the above discussion, there is an argument that can be made: In the case where big data can be made to fuel scientific research and discoveries that are resulting in the amendment of religious texts, can data be considered as the source of ultimate power in terms of governance; thus shattering the age-old conviction that religion holds power? Some researchers seem keen to agree to this proposition while others are not yet convinced that this could be the case. Proponents of data use like Keenan (2018), Anisetti et al. (2018) and Ulbricht and von Grafenstein (2016), among others, points to issues like real-time solutions to problems, efficiency, reduced human interventions and proper planning capabilities in terms of infrastructure, the environment, economy and resource use (Hao et al. 2015). On the contrary, those who still have reservations raises the issues of erosion of democracy, erosion of morals and ethics, data security and commercialization of the public property, as most of the data are handled and controlled by third parties (Fuller 2017; Lake 2017; Kitchin 2016). But these negatives are surmountable and are not anywhere less than those associated with religiously oriented governance structures, as outlined in Chapter 1.

With the use of data gaining traction in governance, a psychological shift of power from religion to data is then deemed inevitable. In particular, this is so since the use of data has been seen to give answers to many uncertainties and give hope for more responsive, safe and sustainable living environments, especially the urban areas. This shift will be made to allow people to interrogate some of the long-held beliefs that have contributed to governance challenges, which have for long resulted into abuse of power, insufficient and inefficient provision of services and widespread social problems. However, can data be one day equated to the positives of region in its way of providing a psychological sense of wellbeing? This will have to be debated in length and deserves a monograph on its own, but surely the advances in Artificial Intelligence and Machine learning may lead towards this.

4.4 Urban Sustainability in the Age of Secularism

The review of literature above has shown that, in the modern era, the imposition of religious dictates in the governance of states and urban areas cannot work. Becci et al. (2016) explain this further by pointing how cities are now home to an incredibly diverse population that pays allegiance to diverse religious or non-religious establishments. In addition, Sezer (2018) notes that most urban centres are, in modern days, a melting pot of cultures and religions, hence, making it a daunting task to pinpoint the dominant religion(s) that would have its precepts in the governing activities of the city, while other religions follow suit piously. Furthermore, Harries et al. (2018) express that urban cities are no longer occupied by indigenous people, hosts people from all walks of lives, and in most cases, each group remain faithful to their religious conviction; hence, imposing religious dogmas of a particular outfit would only make governance daunting and inconsistence. Borrowing from Nieuwenhuis (2012), such a move would be deemed as curtailing freedom of worship, which is inherently provided in most cities and countries. In line to these observations, the separation of religion, state and urban matter is deemed as paramount to lead to more effective governance structures unbiased by any religious dogmas.

The said separation can be considered as already active in most urban areas with secular governance structures being the common trend even at national scale. On this, it is argued that most states have endeavors to maintain a secular profile, with those still under religious dogmas striving to change the tune (Batalla and Baring 2019; Norris and Inglehart 2004). The reason for this is because, most secular states, are seen to work pretty well and offer the ability to act fast on policies, and a majority of them have been able to improve on almost all aspects of governance (Blancke 2018; Fox 2015). Turner (2011) explores this form of governance and argues that it gives space to youth spirituality and a voice to feminism discourse among other modern discourses; hence, opening space for a free society where each group is able to contribute to the betterment of the society. Levey and Modood (2009) posit that is possible since states are governed by constitutions and rule of law, which perceives everyone as being equals, and they provide and protect rights of individuals as well as those of religions; hence, allowing multi-culture and multi-faith within the same locality. An example of secular form of governance includes France, which has even gone further and banned any form of religious symbols in public schools including schools (Judge 2004; Sciolino 2004). However, the extent to which they have gone may also be debated as questions arise if those are truly inclusive and respective of everyone's faith (Laborde 2012; Chrisafis 2013). The United States, and other advanced economies, is seen to lean more towards this secular form of governance, and it would be safe to argue that this has given them a comparative advantage over religious states.

From a scientific point of view, secular societies are better placed and equipped to address climate change, since they are able to come up with unbiased, and effective sustainability policies. This argument is affirmed by Schuman et al. (2018) who found out that religious-based faiths hold a multiplicity of beliefs in regards climate change, with some quarters claiming that it is an act of god. Contrary to this, in the secular world, it is clear that climate change is as a result of human action and/or inaction, and it is only humans that, through proactive policies, can help reversing its effect. On the same, it is evidence that climate change can only be tackled by the adopting of modern technologies and the emphasizing of population controls, among other strategies. The secular world positively embraces, and support policies geared toward such moves, but religions are hesitant on such matters and are seen to be noncommittal in participating in sustainability policies that freely promote such techniques (Poon 2008; Boeck 2014).

In fact, scientific pursuits of dealing with climate change are in perfect harmony with modern ethics and philosophies. This is clear are there is a noticeable change in consumption patterns where the concept of "green" is now highly embraced (Johnstone and Hooper 2016; Codini et al. 2018). The "green" consciousness is even imprinted in corporate and in government work philosophies and these outfits have been observed to go even beyond to support external sustainability initiatives. On the religious front, interestingly enough, issues of sustainability, though mildly emphasized, align perfectly with what many religious texts preach. For instance, most religious texts promote modest consumption and lifestyles, which, in essence, align with the sustainability call. For instance, in Christianity, Jesus is known to preach on consuming what is needed for the day to "give us this day our daily bread". Islam, like many other religions, teaches that the environment is God's creation and should be respected as so (Dien 1997). In Judaism, it is believed that God created man and instructed him to have dominion over the earth and everything therein. Therefore, they are expected to piously protect it, but not to destroy it (Yaffe 2007). Similarly, the clarion calls on most other religions for people to live together in peace and harmony (Banban 2018) has a positive bearing in sustainability drive, as doing contrary to this leads to issues like individualism, which promotes inequalities and uncontrolled consumption, which contributes to unsustainable lifestyles. However, while there are clear indications that sustainability must be pursued by religion, there are contradictions in actions and in certain principles when those are made to adapt to current times. In times requiring urgent climate action, our ethos and principles need to be re-evaluated, even if this means dwelling into sensitive and taboo subjects, and even if this means asking uncomfortable questions.

References

Al-Ghazali AH (2016) Al-Ghazali on the remembrance of death and the afterlife: book xl of the revival of the religious sciences, 2nd edn. The Islamic Texts Society, Cambridge, United Kingdom

Allam MZ (2018) Redefining the smart city: culture. Case Study of Port Louis, Mauritius. Curtin University, Perth, Australia, Metabolism and Governance

Allam Z (2012) Sustainable architecture: utopia or feasible reality? J Biourbanism 2:47–61

Allam Z (2019) Achieving neuroplasticity in artificial neural networks through smart cities. Smart Cities 2(2):118–134

Allam Z, Dhunny ZA (2019) On big data, artificial intelligence and smart cities. Cities 89:80–91. https://doi.org/10.1016/j.cities.2019.01.032

Allam Z, Jones D (2018) Promoting resilience, liveability and sustainability through landscape architectural design: A conceptual framework for Port Louis, Mauritius; a Small Island Developing State. Paper presented at the IFLA World Congress Singapore 2018, Singapore

Allam Z, Newman P (2018) Redefining the smart city: culture, metabolism & governance. Smart Cities 1:4–25

Anisetti M, Ardagna C, Bellandi V, Cremoni M, Frati F, Damaini E (2018) Privacy-aware big data analytics as a service for public health policies in smart cities. Sustain Cities Soc 39:68–77. https://doi.org/10.1016/j.scs.2017.12.019

Ashe G (2001) Encyclopedia of Prophecy. ABC-CLIO, Santa Barbara, California

Ayala FJ (2008) Science, evolution and creationism. PNAS 105(1):3–4. https://doi.org/10.1073/pnas.0711608105

Bačić Ž, Jogun T, Majić I (2018) Integrated sensor systems for smart cities. Tehnički vjesnik 25(1):277–284

Banban D (2018) Harmony in diversity: an empirical study of harmonious co-existence in the multi-ethnic culture of Qinghai. Int J Anthropol Ethnol 2(1) https://doi.org/10.1186/s41257-018-0010-6

Barker DC, Bearce DH (2013) End-times theology, the shadow of the future, and public resistance to addressing global climate change. Polit Res Q 66(2):267–279

Batalla EV, Baring R (2019) Church-state separation and challing issues concerning religion. Religions 10(3):197. https://doi.org/10.3390/rel10030197
BBC (2011) End Times. BBC. http://www.bbc.co.uk/religion/religions/christianity/beliefs/endtimes_1.shtml. Accessed 24th March 2019
Becci I, Burchardt M, Giorda M (2016) Religious super-diversity and spatial strategies in two European cities. Curr Sociol 65(1):73–91. https://doi.org/10.1177/0011392116632030
Blancke S (2018) Islamic modernity and the challenges for secular liberalism. Zygon 53(1):274–287. https://doi.org/10.1111/zygo.12394
Boeck FD (2014) The sacred and the city: modernity, religion, and the urban form in Central Africa. In: Boddy J, Lamberk M (eds) A companion to the anthropology of religion. Wiley
Bose BK (2010) Global warming: energy, environmental pollution, and the impact of power electronics. IEEE Ind Electron Mag 4(1):6–17 doi:https://doi.org/10.1109/MIE.2010.935860
Boyce M (2002) Zoroastrians: their religious beliefs and practices. Routledge, New Fetter Lane, London
Boyer P (2003) Religious thought and behaviour as by-products of brain function. Trends Cogn Sci 7(3):119–124. https://doi.org/10.1016/S1364-6613(03)00031-7
Bryck RL, Fisher PA (2012) Training the Brain: practical applications of neural plasticity from the intersection of cognitive neuroscience, developmental psychology, and prevention science. Am Psychol 67(2):87–100. https://dx.doi.org/10.1037/a0024657
Burkart K (2017) Top fossil fuel companies are driving half of global warming. Leonardo Dicaprio Foundation. https://www.leonardodicaprio.org/top-fossil-fuel-companies-are-driving-half-of-global-warming/. Accessed 25 Mar 2019
Carlisle C (2013) Is religion based on fear? The Guardian. https://www.theguardian.com/commentisfree/belief/2013/dec/02/bertrand-russell-philosopher-religion-fear-christian. Accessed 23th March 2019
Chrisafis A (2013) France's headscarf war: 'It's an attack on freedom'. The Guardian. https://www.theguardian.com/world/2013/jul/22/frances-headscarf-war-attack-on-freedom. Accessed 28 Mar 2019
Codini AP, Miniero G, Bonera M (2018) Why not promote for green consumption?: the controversial role of regulatory focus. Eur Bus Rev 30(5):554–570. https://doi.org/10.1108/EBR-09-2016-0118
Čolaković A, Hadžialić M (2018) Internet of Things (IoT): a review of enabling technologies, challenges, and open research issues. Comput Netw 144:17–39. https://doi.org/10.1016/j.comnet.2018.07.017
Crist E (2007) Beyond the climate crisis: a critique of climate change discourse. Telos 141:29–55
Department of Defense (2015) National security implications of climate-related risks and a changing climate. http://archive.defense.gov/pubs/150724-congressional-report-on-national-implications-of-climate-change.pdf?source=govdelivery. Accessed 22 Mar 2019
Deshpande A, Pitale P, Sanap S (2016) Industrial automation using internet of things (IoT). Int J Adv Res Comput Eng Technol (IJARCET) 5(2):266–269
Dien MI (1997) Islam and the Environment: theory and practice. J Beliefs Values 18(1):47–57. https://doi.org/10.1080/1361767970180106
EPA (2018) Tools for climate change adaptation. Climate Change Adaptation Resource Center (ARC-X). https://www.epa.gov/arc-x/tools-climate-change-adaptation. Accessed 28 Jan 2018
European Commission (2018) 2018-NW Pacific typhoons: past events, current situation and seasonal forecast. European Commission
Faghmous JH, Kumar V (2014) A big data guide to understanding climate change: the case for theory-guided data science. Big Data 1(2):155–163 doi:https://dx.doi.org/10.1089/big.2014.0026
Farber V, Reichert P (2018) Greta Thunberg at Davos: why Gen Z has real power to influence business on climate change. https://theconversation.com/greta-thunberg-at-davos-why-gen-z-has-real-power-to-influence-business-on-climate-change-110409. Accessed 28 Mar 2019
Fox J (2015) Political secularism, religion, and the state. Cambridge University Press, New York, NY

Fuller M (2017) Big data, ethics and religion: new questions from a new science. Religion 8:88. http://dx.doi.org/10.3390/rel8050088
Garwood C (2007) Flat earth: the history of an infamous idea. Mcmillan, London
Goldthau A (2014) Rethinking the governance of energy infrastructure: scale, decentralization and polycentrism. Energy Res Soc Sci 1:134–140
Hao J, Zhu J, Zhong R (2015) The rise of big data on urban studies and planning practices in China: Review and ope research issues. J Urban Manag 4(2):92–124. https://doi.org/10.1016/j.jum.2015.11.002
Harries B, Byrne B, Rhodes J, Wallace S (2018) Diversity in place: narration of diversity in an ethnically mixed, urban area. J Ethic Migr Stud 1–18 https://doi.org/10.1080/1369183x.2018.1480998
Hawking S (2001) The universe in a nutshell. Bantam Press, New York
Henderson RM, Reinert SA, Dekhytar P, Migdal A (2018) Climate change in 2018: implications for business. Harvard Business School 9:317–356
Huang T-J (2017) Imitating the brain with neurocomputer: a "New" way towards artificial general intelligence. Int J Autom Comput 14(5):520–531. https://doi.org/10.1007/s11633-017-1082-y
Hulme M (2009) Why we disagree about climate change: understanding controversy, inaction and opportunity. Cambridge University Press, New York
Johnson TM, Grim BJ (2013) The world's religions in figures: an introduction to international religious demography. Wiley, West Sussex, UK
Johnstone M, Hooper S (2016) Social influence and green consumption behaviour: a need for greater government involvement. J Mark Manag 32(9–10):827–855. https://doi.org/10.1080/0267257X.2016.1189955
Jones RP, Cox D, Navarro-Rivera J (2014) Believers, sympathizers, and skeptics: Why Americans are conflicted about climate change, environmental policy, and science. public religion research institute (PRRI), Washington, DC
Jooste BS, Dokken J-V, Dv Niekerk, Loubser RA (2018) Challenges to belief systems in the context of climate change adaptation. J Disaster Risk Stud 10(1):508
Judge H (2004) The Muslim headscarf and French schools. Am J Educ 111(1):1–24
Karol D (2018) party polarization on environmental issues: toward prospects for change. Niskanen Center
Keenan M (2018) The future of data with the rise of the IoT. RFID J:1–2
Kitchin R (2016) The ethics of smart cities and urban science. Philos Trans R Soc Math Phys Eng Sci 374(2083):115. https://doi.org/10.1098/rsta.2016.0115
Kotenko I, Saenko I, Skorik F, Bushuev S (2015) Neural network approach to forecast the state of the internet of things elements. In: XVIII international conference on soft computing and measurements (SCM), pp 133–135. IEEE
Kumar A, Singh MP, Ghosh S, Anand A (2012) Weather forecasting model using artificial neural network. Proc Technol 4:311–318
Kyle R (2912) Apocalyptic fever: end-time prophecies in modern America. Cascade Books, Eugene, OR
Laborde C (2012) State paternalism and religious dress code. Int J Const Law 10(2):398–410. https://doi.org/10.1093/icon/mor059
Lake RW (2017) Big Data, urban governance, and the ontological politics of hyperindividualism. Big Data Society 4(1) https://doi.org/10.1177/2053951716682537
Leigh I, Hambler A (2014) Religious symbols, conscience, and the rights of others. Oxf J Law Relig 3(1):2–24. https://doi.org/10.1093/ojlr/rwt048
Levey GB, Modood T (2009) Secularism. Cambridge University Press, New York, NY, Religion and Multicultural Citizenship
Levitt D, Kommenda N (2018) Is climate change making hurricanes worse? https://www.theguardian.com/weather/ng-interactive/2018/sep/11/atlantic-hurricanes-are-storms-getting-worse. Accessed 28 Jan 2019

Mahdavinejad MS, Rezvan M, Barekatain M, Adabi P, Barnaghi P, Sheth AP (2018) Machine learning for internet of things data analysis: a survey. Digit Commun Netw 4(3):161–175
Manogaran G, Lopez D (2018) Spatial cumulative sum algorithm with big data analytics for climate change detection. Comput Electr Eng 65:207–221. https://doi.org/10.1016/j.compeleceng.2017.04.006
Markham I (2012) Religion and ethics. Encycl Appl Ethics 748–756. https://doi.org/10.1016/B978-0-12-373932-2.00235-0
NASA (2018) NASA is taking a new look at searching for life beyond Earth. NASA. https://www.nasa.gov/feature/nasa-is-taking-a-new-look-at-searching-for-life-beyond-earth. Accessed 28 Mar 2019
Nath S (2015) Religion and its role in society. IOSR J Humal Soc Sci (IOSR-JHSS) 20(11):82–85
Nieuwenhuis AJ (2012) State and religion, a multidimensional relationship: some comparative law remarks. Int J Const Law 10(1):153–174. https://doi.org/10.1093/icon/mos001
Norris P, Inglehart R (2004) Sacred and secular: religion and politics worldwide. Cambridge University Press, Cambridge
Ogden LE (2018) Climate change, pathogens, and people: the challenges of monitoring a moving target. Bioscience 68(10):733–739. https://doi.org/10.1093/biosci/biy101
Oman D (2018) Introduction to the special issues: spirituality and psychology, emerging perspectives. Pyschol Stud 63(2):89–93
Overstreet DV (2010) Spiritual vs. religious: perspectives from today's undergraduate catholics. J Cathol Educ 14(2):238–263. http://dx.doi.org/10.15365/joce.1402062013
Patel KK, Patel SM (2016) Internet of Things-IoT: definition, characteristics, architecture, enabling technologies, applications and future challenges. Int J Eng Sci Comput 6(5):6122–6131
Poon S-W (2008) Religion, modernity, and urban space: the city god temple in republic Guangzhou. Mod China 34(2):247–275
Reinhart RJ (2018) Global warming age gap: younger Americans most worried. Gallup https://news.gallup.com/poll/234314/global-warming-age-gap-younger-americans-worried.aspx. Accessed 28 Mar 2019
Roser-Renouf C, Maibach E, Leiserowitz A, Rosenthal S (2016) Global warming, god, and the "End Times". Yale University and George Mason University, New Haven, CT, Yale Program on Climate Change Communication
Russell B (1927) Why I am not a Christian. Touchstone Book, New York, NY
Sasmita AO, Kuruvilla J, Ling APK (2018) Harnessing neuroplasticity: modern approaches and clinical future. Int J Neurosci 128(11):1061–1077
Schadewald R (1979) The hollow earth catalogue. TWA Ambassador 12(4):42
Scher S, Messori G (2018) Predictiong weather forecast uncertainty with machine learning. Q J R Meteorol Soc 144(717):2830–2841. https://doi.org/10.1002/qj.3410
Schuman S, Dokken J-V, Dv Niekerk, Loubser RA (2018) Religious beliefs and climate change adaptation: a study of three rural South African communities. J Disaster Risk Stud 10(1):509
Sciolino E (2004) Ban on head scarves takes effect in France. The New York Times. https://www.nytimes.com/2004/09/03/world/europe/ban-on-head-scarves-takes-effect-in-france.html. Accessed 28 Mar 2018
Seidler R, Stevenson RD (2017) Communicating the science and human significance of climate change. Bioscience 67(3):306–309. https://doi.org/10.1093/biosci/biw174
Sezer C (2018) Public life, immigrant amenities and socio-cultural inclusion: the presence and changes of Turkish amenities in Amsterdam. J Urban Des 23(6):823–842. https://doi.org/10.1080/13574809.2018.1475221
Sherwood H (2018) How many believers are there around the world? The Guardian. https://www.theguardian.com/news/2018/aug/27/religion-why-is-faith-growing-and-what-happens-next. Accessed 22 Mar 2019
Short EE, Caminade C, Thomas BN (2017) Climate change contribution to the emergence or re-emergence of parasitic diseases. Infect Dis Res Treatment 10:29. https://dx.doi.org/10.1177%2F1178633617732296

Silton NR, Flannelly KJ, Ellison cG, Galek K, Jacobs MR, Marcum JP, Silton FJ (2011) The association between religious beliefs and practices and end-of-life fears among members of the Presbyterian Church (U.S.A). Rev Religious Res 53(3):357–370

Singh SK, El-Kassar A (2019) Role of big data analytics in developing sustainable capabilities. J Clean Prod 213:1264–1273. https://doi.org/10.1016/j.jclepro.2018.12.199

Skirbekk V, Potančoková M, Hackett C, Stonawski M (2018) Religous affiliation among older age groups worldwide: estimates for 2010 and projections until 2050. J Gerontol 73(8):1439–1445. https://doi.org/10.1093/geronb/gbw144

Sood SK, Sandhu R, Singla K, Chang V (2018) IoT, big data and HPC based smart flood management framework. Sustain Comput Inform Syst 20:102–117. https://doi.org/10.1016/j.suscom.2017.12.001

Stecula DA, Merkley E (2019) Framing climate change: economics, ideology, and uncertainty in American news media content from 1988 to 2014. Front Commun 4. https://doi.org/10.3389/fcomm.2019.00006

Turner BS (2011) Religion and modern society: citizenship. City University of New York, New York, NY, Secularisation and the State

Tzafestas SG (2018) Synergy of IoT and AI in modern society: the robotics and automation case. Robot Autom Eng J 31(5):1–15

Ulbricht L, von Grafenstein M (2016) Big data: big power shifts? Internet Policy Rev 5(1). https://doi.org/10.14763/2016.1.406

UNFCCC (2006) Technologies for adaptation to climate change. UNFCCC, Bonn, Germany

UNFCCC (2018) COP 24 agenda as adopted. UN. https://unfccc.int/node/185180. Accessed 23 Mar 2019

Williams PW (2008) America's religions: from their origins to the 21st century. University of Illinois Press, Urbana-Champaign

Winston A (2019) Young people are leading the way on climate change, and companies need to pay attention. Harvard Business Rev. https://hbr.org/2019/03/young-people-are-leading-the-way-on-climate-change-and-companies-need-to-pay-attention. Accessed 28 Mar 2019

Wright SA, Schultz AE (2018) The rising tide of artificial intelligence and business automation: developing an ethical framework. Bus Horiz 61(6):823–832. https://doi.org/10.1016/j.bushor.2018.07.001

Yaffe MD (2007) Judaism and environment ethics. Philos East West 57(3):400–405

Zorins A, Grabusts P (2015) Artificial neural networks and human brain: survey of improvement possibilities of learning. In: 10th international scientific and practical conference, Rezekne, Latvia, Environment, Technology and Resources, pp 228–231

Chapter 5
Theology, Spirituality, and Urban Objectivity

Abstract The role of religion and urban planning need to be evaluated and explored from an objective standpoint. This must be done void of emotional connections usually accentuated by psychological bias. In this last chapter, the thematic of religion as a political tool is reiterated to support the argument that irrationality can often accompany the subject, and how it could, on the contrary, made to contribute to urbanism from a spiritual standpoint. The works of Christopher Alexander offer some perspective on the matter and further reiterates how design can undermine our innate connection to the universe and made to support religious and spiritual claims. Those are represented through his concept of wholeness, which is supported by other leaders in the field such as Nikos Salingaros. Finally, in the pursuit of objectivity, the thematic of religion and science is evoked and a clear delineation is made as to where each should stand when it comes to urban planning.

Keywords Religion · Urban planning · Spirituality · Wholeness · Christopher Alexander · Politics

5.1 Introduction

As showcased in the previous set of papers, religion can be made to impact on objectivity and digress efforts regarding the input of science in our current societal policies. In particular, Turner (2011) argues that this observation is tied to emotions that religion evokes when some issues that seem to challenge or contradict diverse faiths are proposed or introduced. He argues that those inclined to religion will be up in arms whenever a discovery that seems to challenge their religious beliefs is advanced, and it takes great effort for such to be accepted in the religious domains; often taking decades, if not centuries. Mercier et al. (2018) highlight that the subjective nature of religion is based on the notion of absolute truth held on some issues, and this is even complicated by the fact that such truth is based on a deity that has the absolute authority to change such. The most intriguing issue is that from time past, it has not been possible to separate religion from almost any issue that has the potential to impact on human beings. In this course, Deneulin and Zampini-Davies (2017) explain that even on issues pertaining to development, religion has always played a pivotal role and

Z. Allam, *Theology and Urban Sustainability*, SpringerBriefs in Geography,
https://doi.org/10.1007/978-3-030-29673-5_5

has been observed that the partnership between different developmental stakeholders and faith-based communities have multiplied as time progresses, offering a deeper challenge of achieving objectivity when matters regarding multi-faith are sought.

This being the case, it has been argued that the interrelation that religion has with emotions can create a disruption in the social strata, and on the extreme, be used as a manipulative tool by ill-minded politicians to gain political power. The same can and have been used by religious terrorists and criminals to cause havoc. As cities present themselves as the most densely populated placed in the world, they seem to be the natural target for those attacks, which ultimately impacts on sustainable development-oriented agendas. A case in point is the rise in organized religious terrorist groups that base their argument on particular religious dogmas and teachings and have been forcing these on people through the use of violence. Such groups like ISIS, Al-shabab, Boko Haram and Al-Qaida among many other bases their extremist ideologies on religious believes and these have caused unprecedented harm on cities and economies and can cause larger societal upsets and stress and conflicts between various religious as to their varying definitions of peace.

On a difference lens, religion has also presented significant positive contributions and has the potential to offer incredible impetus especially in regard to social and moral wellbeing. In this respect, it is seen to have the ability to provide the right ingredients for urban planners and also contribute in shaping the efforts of designing better cities. For instance, popular religious cities like Rome and Mecca have been shaped by religious believes dominant in them, and these have been very successful both in terms of designing, economic contribution, especially in regard to religious tourism. Day (2017) explains that even cities that are seen to be more secularized have in one way or the other been inspired by some religious traditions, more so due to the attachment of the diverse citizenry that resides in these cities. She argues that almost every city has traces of religious places of worship, and these have been said to have significant impacts in defining the identities of neighborhoods and activities carried out beyond the boundaries of such neighborhoods.

The works of Christopher Alexander and Nikos Salingaros provides some perspectives on the matter through their exploration on spirituality, wholeness, pattern languages, and resilience. In an article titled *Anti-Architecture and Religion*, Salingaros (2002) comprehensively showcase how religions, since the beginning of time, has influenced building, and the built environment as people tried to express their love for the deity. He explains that the geometrical form of the religious building demonstrated in a perfect way the "wholeness" that a building could evoke, and how that had the potential to influence liveability as people were keen to observe high levels of sanctity to their surroundings (Salingaros 2014). He expresses how religion influenced construction of structurally sound, that depicted high levels of attractiveness, and that evoked the sense of wellness and live to the occupant, and how, through modern architectural designs, such have been rendered worthless. Alexander Christopher on his part, especially in his work, *pattern languages* (Alexander et al. 1977), share the same perspectives with Salingaros, especially in regard to how religion is profound in ensuring that a city survives both through its social strata just as much as in resilient forms. He argues that the religious-inspired buildings have the ability to evoke a sense of living and wholeness, especially in regard to the limitless number of building patterns that can be inspired by this, depicting the uniqueness of the people living in the society and their perception of deity (Alexander 1979). In his work, *The Nature of Order* (Alexander 2002b), he argues that religion can be expressed through the application of different expressions and patterns that are ordered in such a way that they depict a harmonious complexity that promote a sense of wellbeing and reasserts comfort. On this, Salingaros (2000) supports that the said complexity yields a matrix of connectedness that promotes a sense of "wholeness", and when modern architectural trends, inspired by anti-religious themes, are emphasized, the outcomes are amplified by disorganized interaction and movement of people.

However, where there are numerous principles like those of oneness and harmoniousness among many others that can be borrowed from religious texts, but those abstract perceptions should be treated with care. In particular, the texts should not be taken in their literal form and applied in urban design, as some can have negative impacts on liveability dimensions. This is so noting that different religious outfits are inspired by unique, but, sometimes contradicting principles, and if each is taken literary, the "wholeness" and complexity emphasized by Alexander (Seamon 2007; Alexander 2002b) and Salingaros (2000, 2003) would be eroded, and instead result to disorganized and unattractive patterns. Therefore, in our current times, where things are increasingly influenced by scientific perception, the emphasis should be to find a harmonious balance between religion and science in the quest for finding appropriate urban planning tools. This way, it will be possible to accommodate and address trending issues like urbanization and population increase while promoting resiliency, sustainability and liveability in cities, leading to the actualization of the Sustainable Development Goal 11 and the New Urban Agenda.

5.2 Religion and Politics: An Explosive Mixture

The power that religion holds on people due to its innate emotional connection can be a dangerous tool used by political leaders with an aim to secure power at any cost. Buying from the Marxist view of religion, there are numerous, tangible evidences that show that religion has helped create a clique in every society using religious teachings to elevate themselves above the rest in the society, with the sole aim to achieve authority for increasing political influence (Marx and Engels 1975). Silton et al. (2011) point out that such people invoke religious texts to instil a sense of fear and emotions; hence, the masses are unable to question teachings that have led to them being alienated and oppressed. In keeping with this line of thought, Atkinson and Bourrat (2011) explain that fear of questioning the deity has allowed a few who argue that they are directly appointed from above to have total authority, and in most cases, they have used such powers to advance their own selfish agendas. Muthuswamy (2016) expresses that this is even worse when such leaders also hold political offices since they have the opportunity to influence decision-making processes. For this reason, any dissenting voices objectively criticize or offer an alternative view on issues have been seen to be dismissed in manners that make others with similar view to cower and remain silent.

The religious authority yielded by a few, unfortunately, has been seen to perpetuate increased social segregation in different societies, and these have led to unprecedented conflicts and anarchy in different countries. For instance, the perpetual violence between Israel and Palestine is all based on religious convictions, though, with intricate political ends (Lewis 2018; Lintl 2918). Though the issue is contested, the Israelites believe that the area they occupy is an inheritance promise to Abram by God, while the Palestinians view this as an outright encroachment of their land (Maranville 2002). Another example that showcases how religion has influenced violence is the case of the violence perpetuated on the people of Uganda by the Lord Resistance Army; an outfit basing their cruel and violent action on the biblical teachings. The leaders of the outfit, who were out to gain political mileage argued that the country ought to have been guided by religious principles (Mwenda 2010). In Somalia, religion has been used by leaders with political ambition to cause havoc on the entire country, such that, the country is in total disarray. The leaders sponsors terrorist groups like the Al-Shabaab, which disguises as one advancing the need for observance of religious teachings to cause havoc in towns in the country and even in neighboring country; Kenya (Curtis 2019). In Southern Sudan, the violence that has persisted to date, as explained by Basedau and Koos (2015) is seen to be supported by religious leaders who support different political outfits in the country. They argue that this is based on issues like religious intolerance where different religious faiths are seen to support political leaders depending on their religious inclination. Even in developed countries, especially in the recent past where there has been increased concerns on immigrants, political leaders have been seen to exacerbate advance political ideologies that have the potential to spur religiously inclined violence (Gleditsch and Rudolfsen 2016). This is more so due to religious profiling. Such trends can even be attributed to the case of terrorism where a Caucasian young man killed dozens of people in New Zealand due to varying religious inclination (Van Sant 2019).

It is unsurprising that most of the aforementioned conflicts happen in cities, and this can be attributed to the fact that this is where the density of people is highest (Robb 2007). Also, cities are host to numerous, critical political, economic and social establishment that when attacked, would have far-reaching political implications and increased attention (Graham 2004). Burke (2018) explained that when violence happens in a city, regardless of its magnitude, those have the potential to attract a lot of attention; hence, in one way or the other spark a discourse that may cause a political or public opinion, which often helps to market the groups' ideologies and attract more people to their cause. Goodfellow (2017) explains that cities also carry the greatest influence on the political future of any country and thus, the ill-minded political leaders would want to disrupt the order of cities to allow them to continue holding on power, or gain power.

Unfortunately, the manifestations of this violence, especially those with religious inclinations, principally negate the principles brought by religion more than they tie people to it. This is demonstrated in the examples cited above, where, in each particular case when violence has erupted, people have been left more divided, and political leaders who instigated or sponsored the violence are left better off than the masses directly involved in those.

5.3 Spiritual Potency of Religion in Urban Places

Instead of fighting towards segregation, religion and the power it yields should be used to bring forth a culture of inclusivity and oneness. Indeed, the writings of Fagan (1996) underline that different religions have been used as tools for the unification of people from different backgrounds in a common goal of worshiping the deity. In this regard, Burch-Brown and Baker (2016) acknowledge that religion has been used in many forums, and is seen as among the most successful tools for fighting and reducing such issues like prejudice, racial profiling, fighting social discrimination, and also in advocating equity in the society. For instance, in the case of apartheid in South Africa, Smock (2015) reports that the coming together of different religions groups played a critical role in its defeat and also in ensuring that a peaceful transition was achieved. A similar success was achieved in Mozambique in 1992, when a Rome-based Community of Sant'Egidio was able to negotiate peace; leading to the end of civil war in the country, that saw the deaths of between 600,000 and 1 million people during a 16 year period lasting between 1976 and 1992 (Mass Atrocity Endings 2015).

The main drive towards inclusivity is for various religious groups is to look at the spirituality aspects of religion, since the political, social and economic aspects of it have been proven to accelerate segregation, alienation and other religious-based social problems. McKinnon (2017) argues that by looking at religion through a social lens alone, it may be argued to have failed since it promotes creation of different hierarchies that are in themselves discriminative and not inclusive. This is demonstrated even in such sectors like education and health where religion has invested heavily in the access to those services by various social classes, assigned by religious affiliations. Indeed, according to Thompson (2015), these sectors portrays outright discrimination and segregation based on inter-religious backgrounds. In the education sector, the authors argue that faith-based schools were breeding grounds for fundamentalists and extremists, and such types of school only help in expanding the social boundaries created by religion. In particular, this is magnified when such policies regarding single faith-based schools are backed by government policies, especially where there exist minority religious outfits, hence, supporting the outset that the government is supporting selected dominant religious groups.

In the case of spirituality, the goal of each religion is for individuals to acquire deeper meaning and connection with the deity, and in this, all religions have a convergent view. This is particularly demonstrated in the current trends with the youths who are increasingly denouncing forms of relationships with religious groups but are deeply ingrained in spirituality and all it has to offer. Turner (2011) argues that spirituality has provided an opportunity for the youth to freely express themselves and as such, they are able to avoid unnecessary religious confrontation that has been advanced by traditional religious thinking. Overstreet (2010) stresses that religion is not seen to have answers to a myriad of questions that confront the society, and this has prompted the youths, who are seeking such answers to dissociate with religion, and by doing so, they are able to freely interact, without the religious walls that, as explained above, have prompted numerous social, political and economic disorders.

In urban areas, the principles of spirituality can be harnessed and explored in the designing processes so as to create a harmonious and inclusive built environment as explained by Alexander et al. (1977) and Salingaros (2000). In particular, this can be incorporated in urban regenerative efforts, especially in the public urban spaces that serve as meeting points for people from all broads of life. Like the spirituality aspects that allow people to rise above their conservative religious beliefs, public urban spaces ought to offer serene environments that promote the qualities of oneness, and where everyone is free to converge. This idea of borrowing from spirituality is advanced by De Beer (2018), who argues that that the same concept can be utilized to check on issues like gentrification that are eminent when urban revitalization takes place. According to him, the idea put forward by Roberts (2000) of ensuring regeneration focuses only on the physical upgrading, is valid since, at the end of the process, everyone, including the poor who may be relocated get an opportunity to also contribute to the environment. On the contrary, when this is not observed, he argues that the regeneration would take the same form as the negative aspects of religion; that of social stratification and alienation.

On the same, noting that most ancient religious cities valued and used public spaces for numerous purposes, such were fashioned to emphasize the religious orientation the dominant religion of the time. But, in the modern age where religion is on the decline and is being replaced by spirituality, such public spaces need to be realigned to correspond with the changing philosophies represented by spirituality. While on this, such need to be accommodative of a larger and more diverse audience, while at the same time, ensure that some of the geometry that allows for harmony is not replaced by the monotonous designs that Salingaros (2002) warns about.

5.4 Achieving Wholeness

When talking about achieving inclusivity and building coherence through the study of religious principles, the works of Christopher Alexander brings forth important contributions to the discussion. Throughout his works, as is summarized by Dawes and Ostwald (2017), he explains that his desire was to articulate, through design, art, and building, the uniqueness and wholeness and aliveness that is demonstrated in the universe and how delicately, yet complex it was fashioned. The emphasis on Alexander's work is vested in the need for each design, regardless of its size or magnitude to achieve the objective of preserving the nature of order, bring out liveliness that would invoke a sense of comfort and harmony on those who consumed his works and ensure it promoted deeper interconnectedness. Like in religion, he has endeavored to demonstrate that the built environment is not just a matter of having scattered works of outer beauty, but how much the collection of those could help to foster an environment aimed to help those living in it to find inner fulfilment and wellbeing. Such quests have been demonstrated to be alive as people from all walks of life have been observed to be in search for deeper spiritual understandings of the supernatural, and have demonstrated this through the type of religious places of worship they built. Salingaros (2002) explains that every aspect of such buildings, were done to exemplify the nature of God and mystic forms; hence, the uniqueness of each building that each religious group builds. Similarly, Alexander has tried to bring out this notion that each tangible substance of architecture has the ability to point out to the reality of God, and such, should be aligned to assist those who interact with it, in one way or the other, experience wholeness, comprehensively addressed in his book series, entitled the "*Nature of Order*" (Alexander 2002a, c, d, e).

In his book, *The Luminous Ground (Alexander)*, Alexander tries to demystify how the nature of matter as we know it is not machine-like; in line with most of the contemporary science. According to him, the human pursuit of happiness, fulfilment and inner understanding of the universe through science only plunge them into deeper despair and hopelessness, as they are made to conclude that they and the matter that makes nature is just like a machine. In this case, he argues that everything for those who subscribe to this school of thought try is to comply with mechanistic qualities aligned with science. The reality that Alexander advocates are that neither human beings, nor the matter in nature is mechanistic and the type of built environment that follows the scientific perception of the universe cannot suffice, at any cost, help people relate with the universe, or find comfort or liveliness in them. According to him, designs inspired by the scientific world-picture, at best, only succeed in producing structures that are inert, that, though they demonstrate intricate use of geometrical substance, are void of life-in effect, and those who live in them can only be hopeless. He argues that until people acknowledge that the universe and all matter in it are not neutral and inert, but alive and conscious can they then manage to construct a built environment that is full of life, that reverberates with nature and that is worth living in. For instance, he explains how the use of the right color, can help bring the sense of inner light, which, in a deeper meaning amplifies the phenomena of life, and subsequently create a sense of liveliness that human being so earnestly yearns to achieve.

Alexander's conviction about how architectural designs ought to bring the best of life in every human is also shared by Salingaros (2002), who argues that buildings should be emotionally-nourishing. Like Alexander, he argues that the modern trends of designing built environments that only exacerbate strive, loneliness and ill-feeling, are deliberate, as people try to run away from the fact that human being and nature are interconnected, and both are living. Such structures, according to him are inspired by mysticism and irrationality, which Alexander views as being mechanistic pursuits. Salingaros and Kalb (2012) explain that the most unfortunate thing, in this case, is that modern architects and designers are aware of the traditional and most fulfilling processes that can be applied to achieve a living environment, but, they choose to follow the wider path that leads to vanity. Such vanity, they argue is demonstrated in the modern buildings that are less functional, less adaptive and less human. According to them, as was also the conviction of Alexander in his book *Luminous Ground*, the failure of modern architects in rejecting the fact that a living, built environment can only be achieved when there is deliberate desire to be inspired by both the humanistic and religious mind-set, as was the case in the traditional era (Salingaros 2019). In such a case, the designs that proceed from such laborious works have no room to allow human beings, who are the target consumers of such to link their deeper self with this living environment. Therefore, just like in religion void of spirituality, these will remain disconnected from their environment, which, in reality, need to be a place of harmony, comfort, and wholeness.

5.5 Religion, Science and the Shaping of the Urban Realm

The dimensions of inclusivity and wholeness advocated in this chapter also need to expand from the boundaries of religion to acknowledge the role of science, which, in the twenty-first century cannot be ignored exclusively in favor of religious subjectivity. In the above discussions, the call was to ensure that science does not override the role of religion and spirituality by treating every element of the urban fabric, including people therein as machines that have no feelings, no conscious or that are not living. Similarly, it was earlier observed that religion canons cannot alone be used to guide the designing of urban centers, especially with respect to regeneration (Patitsas 2010). Therefore, for this reason, as was advanced by Jacobs (1961), who vehemently objected the urban renewal of American cities, if the plights of those living in them were taken into consideration. According to her and just like Alexander argued, cities everywhere should be treated more like biological organisms that have the potential to die if treated in a disorganized complexity. In her view, the planned renewal of American cities that was to begin in 1949 after the United States Congress passed the Urban Renewal Act permitting the renewal of cities originally made to comply with modern trends. Consciously, the planners involved in this exercise had deliberately planned to phase out all the traditional aspects of cities and replaced them with the modern trends of the day, illuminated by gleaming architectural structures. Jacobs argued that this meant killing the intricate webs of intimate human relationship and the complex commercial networks that characterized those cities; as supported by Salingaros above. Through great effort, in 1974, Jacobs and her supporters managed to drum sense in the need to address renewal programs observing the sanctity and serenity of streets, mixed-use neighborhoods and the active and stable social networks characterized in cities.

The need for the balanced use of science and religion is also advocated by Salingaros (1999), who argues that there exists an intimate relationship between architecture and mathematics and a combination of these two allow for harmonious architectural patterns. He explains that mathematics is a science of patterns, and such patterns are paramount in the built environment, as they help designers achieve the concept of wholeness and inclusivity. In the article, *Anti-Architecture and Religion* (Salingaros 2002), he showcased how religion inspires structures that promote inclusivity and wholeness, and here he also explains how the frictionless relationship between these religion and science can help achieve even better results. He argues that this was piously observed until the twentieth century when the modern architectural designs started to be fashioned void of mathematical influence; hence, the concept of patterns was rendered obsolete but this only helped create monotonous designs. In the article titled, *Architecture, Patterns, and Mathematics*, Salingaros (1999) have showcased that patterns play a critical role in architectural design, especially if such are used to solve issues with factors like space. According to him, disregarding such importance degenerates into having structures that only generates negative emotional responses for lack of human qualities. The knowledge of these patterns rendered by mathematical application in structures is the same advanced by Alexander in his work, *Pattern Language* (Alexander et al. 1977) where he explains that when patterns lead to a solution and this is repeated in time and space, the outcomes render a built environment that oozes with liveliness and that promotes inclusivity, as those who live in them are able to experience inner peace and harmony, and they are able to replicate this on others. Such is the case of the traditional religiously inspired structures that allowed people to connect even better with nature, and by so doing, were able to live in harmony.

The idea of favoring either religion or science while designing modern cities is perceived as a precursor to creating emptiness, and Salingaros (1999) explains that this does not have any content, this leads to chaotic and disorganized built environments that cannot promote human coexistence. Nevertheless, when the two concepts are harmoniously balanced, they promote the creation of structures that obeys the natural orders as well as promote the application of modern innovation like the sensitive use of technology. On the same vein, Klinger and Salingaros (2000) highlight that a coherent application of both science and religious dictates have the ability to allow for designing of structures that achieve complexity and makes use of pattern languages, and these, in turn, offer a large quantity of information that is well ordered and coherent such that, the mind of those living in them are able to comprehend and relate well with nature and the supernatural.

In sum, urban planners have an ethical and moral obligation to design for people and thus we cannot deal in abstraction, which is usually favored by religious leaders. Similarly, they should not also succumb to the dictates of modernists who promote the idea of science, without regard to the emotional of the human and the living character of the urban fabric they fashion. Therefore, as urban planning is the science for designing for humanity, planners have the moral and ethical imperative to consider all the tools at their disposal; being either religious and scientific to ensure that cities are functional, liveable and promote sustainability in all aspects.

References

Alexander C (1979) The timeless way of building. Oxford University Press, New York

Alexander C (2002a) The luminous ground: an essay on the art of building and the nature of the universe. In: Alexander CWJ (ed) Essay on the art of building and the nature of the universe. Center for Environmental Structure, Berkeley, Calif
Alexander C (2002b) The nature of order: the process of creating life. The Centre for Environmental Structure, Berkeley, California
Alexander C (2002c) The phenomenon of life: an essay on the art of building and the nature of the universe. In: Alexander CWJ (ed) Essay on the art of building and the nature of the universe. Center for Environmental Structure, Berkeley, Calif
Alexander C (2002d) The process of creating life: an essay on the art of building and the nature of the universe. In: Alexander CWJ (ed) Essay on the art of building and the nature of the universe. Center for Environmental Structure, Berkeley, California
Alexander C (2002e) A vision of a living world: an essay on the art of building and the nature of the universe. In: Alexander CWJ (ed) Essay on the art of building and the nature of the Universe. Center for Environmental Structure, Berkeley, California
Alexander C (2004) The Luminous ground. The Center for Environmental Structure, Berkeley, California
Alexander C, Ishikawa S, Silverstein M (1977) A pattern language. Oxford University Press, New York
Atkinson QD, Bourrat P (2011) Beliefs about God, the afterlife and morality support the role of supernatural policing in human cooperation. Evol Hum Behav 32:41–49
Basedau M, Koos C (2015) When do religious leaders support faith-based violence? Evidence from a survey poll in South Sudan. Polit Res Q 68(4):760–772
Burch-Brown J, Baker W (2016) Religion and reducing prejudice. Group Process Intergroup Relat 19(6):784–807. https://doi.org/10.1177/1368430216629566
Burke J (2018) Cities and terror: an indivisible brutal relationship. https://www.theguardian.com/cities/2018/jan/29/cities-terror-attacks-brutal-relationship-terrorism. Accessed 16 June 2019
Curtis B (2019) Al-Shabaab's Kenya attack proves the terrorist group is still deadly. Vox. https://www.vox.com/world/2019/1/16/18185182/nairobi-kenya-hotel-attack-spindler-american. Accessed 12 June 2019
Dawes MJ, Ostwald MJ (2017) Christopher Alexander's a pattern language: analysing, mapping and classifying the critical response. City Territ Arch 4(17):1–14
Day K (2017) Urban space and religion in the United States. Oxf Res Encycl. https://doi.org/10.1093/acrefore/9780199340378.013.470
De Beer SF (2018) Faith-based action and urban regeneration. UTS Theol Stud 74(3):1–11
Deneulin S, Zampini-Davies A (2017) Engaging development and religion: methodological groundings. World Dev 99:110–121. https://doi.org/10.1016/j.worlddev.2017.07.014
Fagan P (1996) Why religion matters: the impact of religious practice on social stability. The Heritage Foundation, Washington, DC
Gleditsch NP, Rudolfsen I (2016) Are Muslim countries more prone to violence? Res Polit 3(2):1–9. https://doi.org/10.1177/2053168016646392
Goodfellow T (2017) Seeing political settlements through the city: a framework for comparative analysis of urban transformation. Dev Chang 49(1):199–222. 0.1111/dech.12361
Graham S (2004) Cities, war, and terror. Blackwell, Malden, MA
Jacobs J (1961) The death and life of great american cities. Random House, New York
Klinger A, Salingaros NA (2000) A pattern measure. Environ Plann B Plann Design 27:537–547
Lewis KL (2018) Peace in israel and palestine: moving from conversation to implementation of two-state solution. Southwest J Int Law 24:251–274
Lintl P (2018) Actors in the Israel-Palestinian conflict: interests, narratives and the reciprocal effects of the occupation, vol 3. German Institute for International and Security Affairs
Maranville C (2002) Palestine Israel whose land is it, really. Beyond Today. https://www.ucg.org/world-news-and-prophecy/palestine-israel-whose-land-is-it-really
Marx K, Engels F (1975) On religion. Progress Publishers, Moscow

Mass Atrocity Endings (2015) Mozambique: civil war. Mass Atrocity Endings. https://sites.tufts.edu/atrocityendings/2015/08/07/mozambique-civil-war/#_ednref21. Accessed 17 June 2019
McKinnon A (2017) Religion and social class: theory and method after Bourdieu. Soc Res Online 22(1):1–13. https://doi.org/10.5153/sro.4247
Mercier B, Kramer SR, Shariff AF (2018) Belief in god: why people believe, and why they don't. Curr Dir Psychol Sci 27(4). https://doi.org/10.1177/0963721418754491
Muthuswamy MS (2016) The role of Sharia and religion leaders in influencing violent radicalism. Sci Relig Cult 3(1):1–18
Mwenda A (2010) Uganda's politics of foreign aid and violent conflict: the political uses of the LRA rebellion. In: Allens T, Vlassenroot K (eds) The lord's resistance army: myth and reality. Zed Books, London, pp 45–58
Overstreet DV (2010) Spiritual vs. religious: perspectives from today's undergraduate catholics. J Cathol Educat 14(2):238–263. http://dx.doi.org/10.15365/joce.1402062013
Patitsas T (2010) The city as liturgy: an orthodox theologian corresponds with Jane Jacobs about a gentle reconciliation of science and religion. New Rome Press
Robb J (2007) The coming urban terror: systems disruption, networked gangs, and bioweapons. CJ. https://www.city-journal.org/html/coming-urban-terror-13026.html. Accessed 15 June 2019
Roberts P (2000) The evolution, definition and purpose of urban regeneration. In: Roberts P, Sykes H (eds) Urban regeneration: a handbook. Sage, London, pp 9–36
Salingaros NA (1999) Architecture, patterns, and mathematics. Nexus Netw J 1(2):75–86
Salingaros NA (2000) Complexity and urban coherence. J Urban Des 5:291–316
Salingaros NA (2002) Anti-architecture and religion. Catholic Culture. https://www.catholicculture.org/culture/library/view.cfm?id=4736. Accessed 16 June 2019
Salingaros NA (2003) Connecting the Fractal City. In: Paper presented at the 5th Biennial of towns and town planners in Europe, Barcelona, April 2003
Salingaros NA (2014) Complexity in architecture and design. Oz J 36:18–25
Salingaros NA (2019) Beauty and the nature of matter: the legacy of Christopher Alexander. New English Rev. https://www.newenglishreview.org/custpage.cfm?frm=189673&sec_id=189673 Accessed 16 June 2019
Salingaros NA, Kalb J (2012) Why contemporary architecture is against god and man. Crisis Magazine. https://www.crisismagazine.com/2012/why-contemporary-architecture-is-against-god-and-man. Accessed 18th June 2019
Seamon D (2007) Christopher Alexander and a phenomenology of wholeness. In: Paper presented at the Christopher Alexander annual meeting of the Environmental Design Research Association, Sacramento, CA
Silton NR, Flannelly KJ, Ellison CG, Galek K, Jacobs MR, Marcum JP, Silton FJ (2011) The association between religious beliefs and practices and end-of-life fears among members of the Presbyterian Church (USA). Rev Religious Res 53(3):357–370
Smock DR (2015) Religious contributions to peacemaking: when religion brings peace. Not War, United States Institute of Peace, Washington, D.C.
Thompson D (2015) Banning faith schools is no quick fix to social segregation. The Conversation. http://theconversation.com/banning-faith-schools-is-no-quick-fix-to-social-segregation-38797. Accessed 16 June 2019
Turner BS (2011) Religion and modern society: citizenship. City University of New York, New York, NY, Secularisation and the State
Van Sant S (2019) Accused shooter in New Zealand mosque attacks charged with terrorism. NPR. https://www.npr.org/2019/05/21/725390449/accused-shooter-in-new-zealand-mosque-attacks-charged-with-terrorism. Accessed 14 June 2019

Conclusion

This book presents literature on how religion has, since ancient days, been present and dominant in governance of states and even urban areas, and how we are witnessing it is losing its footing. It has demonstrated that in the modern days, some countries, states, and urban areas are more inclined to adopting secular governance structures in favor of those pegged on religious dogmas. In particular, they have been influenced by the perception that religion has been laxed in pressing issues like climate change that require speedy and concrete action. For instance, the stance of different religions in regard to practices like matrimony and disposal of the deceased can be showcased as being counterproductive to both planning ideologies from environmental perspectives. In regard to religious matrimony, the emphasis for each household, for instance, to have a shelter with lawns and gardens has been accused of fueling sprawl, which has had negative impacts on the environment. Similarly, the insistence of religion for families to have many children is also seen to be against the principles of sustainability, as this promotes population increase and incidentally an equal increase in consumption of resources among many other things.

On the issue of resting the dead, different religions, especially those that advocate for burials are said to contribute greatly in graveyard sprawl in different urban areas across the globe. The aftermath of this affects both societal liveability standards and sustainability drives. On this, pro-sustainability advocates are said to persuade religions to adopt and emphasize on such practices like cremation and grave reuse, without however compromising the dignity of the deceased. By so doing, it is argued that such approaches would facilitate the conservation of environmentally sensitive areas and maintain control on sprawl (both urban and graveyard), while at the same time ensuring the safety, comfort and liveability perception of neighborhoods surrounding graveyards. Similarly, such strategies would allow for more adaptive reuse of those spaces; such as affordable housing or green open spaces and lead to more healthy lifestyles.

Despite what may have been perceived as wholesomely negative on the matter of religions, which are by a higher margin debatable, it should not be lost that religion also has been playing significant roles in ensuring that sustainability prevails. From the literature surveyed in this book, it was discussed that a majority of religions

Z. Allam, *Theology and Urban Sustainability*, SpringerBriefs in Geography,
https://doi.org/10.1007/978-3-030-29673-5

have been in the forefront in encouraging and emphasizing the need for modest living; hence, promoting subtle use of resources. As they may be argued as being the foundation of social structures, religions are also credited for being in the fore in encouraging peaceful and harmonious coexistence with nature and between human beings. And luckily, the association of religion with deities equally means that all the precepts pronounced are bound to be followed piously; thus, the harmonious call is not established in poetic (and cryptic) verses but can be seen to be observed faithfully, even though that religiosity is on the decline. On this still, most religious organizations have been observed to promote such initiatives like education, healthcare, protection of human right and resettlement programs, which are all key drivers in ensuring sustainability in all levels; social, economy and environment. Due to its influence, one cannot help to infer that there is a common link between religion and the political class; which is, in most cases tasked with the mandate of spearheading sustainability drive through policy formulation, implementation of the rules of the law and legislation laws. Religions, on their part, have been very vocal on matters relating to governance; hence, their emphasis on the need for policies that are deemed important by their adherents.

While the principles of sustainability as promoted by religion are not disputed in this work, it is argued that religion must be kept disparate from state and urban affairs and be made to operate from sound principles and scientific data. By doing this, such cardinal principles like social inclusivity, security, economic growth, sustainability, and resilience would be promoted at greater lengths. This proposition is based on the literature review outlined in this book where it was established that a majority of urban areas are now seen as centers for multi-culture and multi-religions, hence, teachings from a singular religious group cannot be made to apply for everyone, especially noting that religions differ principally on ideologies.

By separating religion from state governance, it has been observed that the society can be made to enjoy even larger spaces of freedom of expression and worship, which is better for urban areas synonymous with high populations and diversity of cultures. With these, even in cases of discussion pertaining to different issues confronting urban areas or states, it has been argued that stakeholders can manage to engage in sober, extensive, inclusive and productive discourses based on logic, experience, and data, as no one is bound by stringent religious or state rules. This is possible since each entity is independent from each other, thus, no fear of losing endorsement, like is the case where separation is absent, and groups are not keen to offend each other to avoid being ostracized in future discussions.

On a different note, without religion or the state interfering with the affairs of each other, it would be possible to institute leadership based on capabilities and qualifications that could be based on merit backed by scientific data. With such leaders, in the case of cities, transformative agendas like implementation of smart city concepts or specialized technologies would be fast-tracked; thus, enabling the basis of building a city that will be able to source data from its residents for sound and intelligent planning and governance. On the religious front, they would manage to advance their faith and achieve their core mandate, as separation mean that each religious group would have equal opportunities and right to conduct their affairs as long as they do not contravene the laws of the land.

The above propositions, however, are tricky to institute as religion has in the past influenced state affairs, laws and legislations, and these are deeply engrained in our societal life; hence, the need to identify those are important in order to better re-evaluate what needs to be addressed to achieve a true separation of religion and state. The pursuit here is not to be antagonistic to religion, but to adopt an approach of philosophical critical thinking that can help to re-evaluate our ethos towards sustainability, something that paradoxically aligns with most of what religion teaches. Using force and underhand tactics to persuade religion to separate itself from state affairs would not work, as different religious groups have complex and widespread networks, that, when triggered, can only frustrate any meaningful urban agenda from taking place. Consequently, on this, it is worth noting, as is observed in literature, that a majority of those in position of influence in the society portrays, in one way or the other some religious persuasions; hence, any effort that would be deemed to undermine their faith would be considered as advance defeat. This is true as a majority of those in leadership, especially in less developed countries have been found to attribute their position with anointing and endorsement from a combination of higher political and religious authorities, and a disruption in their way of thought can lead to detrimental consequences on the city and the peace and inclusivity that it must uphold in order to achieve sustainability. A new (gradient) approach is thus required equating both religious principles and supported by science. The works of Christopher Alexander seems to lead to this new way of understanding and can provide the blueprint for regenerating cities.

Ensuring that the adopting of philosophical critical thinking would also allow religion to maintain its critical role in the society while including sound governance structures. This will ensure that those in other realms would be free to perform their duties freely without fear of being policed by religious or state's dictates. By so doing, even radical approaches, deemed secular in nature, aimed towards the achievement of sustainability could be made supported by religion; as those will be ultimately geared towards uplifting societal standards and lead to more diverse and vibrant cities.

This book is in no way an attack on religion and the values it upholds. Instead, it seeks to underline that the ethos highlighted by theology must be re-evaluated and uplifted to ensure that religious dogmas are truly adaptive to contemporary societies and to current climate challenges that our world faces; without which, humanity may face a deeper state of crisis.

MIX
Papier aus verantwortungsvollen Quellen
Paper from responsible sources
FSC® C105338

If you have any concerns about our products,
you can contact us on
ProductSafety@springernature.com

In case Publisher is established outside the EU,
the EU authorized representative is:
Springer Nature Customer Service Center GmbH
Europaplatz 3, 69115 Heidelberg, Germany

Printed by Libri Plureos GmbH
in Hamburg, Germany